The Consciousness Revolutions

Shimon Edelman

The Consciousness Revolutions

From Amoeba Awareness to Human Emancipation

Springer

Shimon Edelman
Department of Psychology
Cornell University
Ithaca, NY, USA

ISBN 978-3-031-24011-9 ISBN 978-3-031-24012-6 (eBook)
https://doi.org/10.1007/978-3-031-24012-6

This Springer imprint is published by the registered company Springer Nature Switzerland AG
The registered company address is: Gewerbestrasse 11, 6330 Cham, Switzerland

TO THE MEMORY OF URSULA K. LE GUIN

"I'll make my report as if I told a story, for I was taught as a child on my homeworld that Truth is a matter of imagination."

— *The Left Hand of Darkness*

Acknowledgments

The "revolutions" in the title of this book is an echo of Szathmary and Maynard-Smith's celebrated 1995 paper "The major evolutionary transitions." Neither the book nor any of my forays into consciousness research would have happened if it were not for Tomer Fekete, whose work on the intrinsic topology of cognitive representations led to many discussions and eventual collaboration. My thinking about consciousness has been greatly influenced by the work of Daniel Dennett, Björn Merker, and Thomas Metzinger. Over the years, I have benefitted from talking about consciousness, construed broadly (as it is here), with Aman Agarwal, Carsten Allefeld, Leonor Blum, Matt Botvinick, Björn Brembs, Dave Chalmers, Antonio Chella, Paul Cisek, Austen Clark, Rick Dale, Eric Dietrich, Hetvi Doshi, David Edelman, Steven Elmlinger, Barb Finlay, Karl Friston, Joe Halpern, Stevan Harnad, John-Dylan Hayes, Oren Kolodny, Tamar Kushnir, Jakub Limanowski, Dan Lloyd, Arnon Lotem, Malcolm MacIver, Rafi Malach, George Mashour, Björn Merker, Thomas Metzinger, Melanie Mitchell, Owen Morgan, Roy Moyal, Victor Odouard, Steven Piantadosi, Saeedeh Sadeghi, Harry Segal, Camilo Signorelli, Amritpal Singh, Lee Smolin, Michael Spivey, Marius Usher, Cees van Leeuwen, Anne Warlaumont, Bryan West, Neta Zach, Yi Zeng, and Vivian Zhang. Special thanks are due to Aliou Gambrel, Emma Plowe, Deanndria Singleton, and the rest of the participants in my consciousness seminar at Cornell in Spring 2022, for many illuminating comments on the readings, which included early drafts of parts of the present book. Eric Dietrich, Karl Friston, Tom Gilovich, Luca Onnis, and Emma Plowe were kind enough to read and comment on all or parts of the penultimate draft.

As always, I am grateful to my family for continued encouragement and support. Lastly, I take this opportunity to thank my father, for having acquired, a long time ago, a copy of Y. L. Fung's *A Short History of Chinese Philosophy* (1948), which concludes with a sentiment that I find more and more appealing as I grow older:

> One must speak very much before one keeps silent.

Contents

Part I The Human Condition 1

Prelude 3

The story of Oh: a brief account of consciousness, from Amoeba to Zoloft.

1 Selfless Consciousness 7

Tower of power. The brain's virtual reality. Identity theory. Computation all the way down. No cognition without representation. What makes representation special and how it can give rise to basic phenomenal awareness. Stimulus and response and awareness and zombies. Slipping into the future. Once again, with feeling. The Gang of Four. Minimal consciousness. The first revolution.

Notes 23

References 28

2 Minimal Selves 35

Tired starlings. The web of cause and effect. Emergence: the effective self as a local causal domain. Emergence and causality: Where has all the physics gone? The self and others. The captive scapegoat and the birth of tragedy. The second revolution.

Notes 43

References 48

3 Self-Consciousness 53

Self-affirmation. Self-evidencing. Self-creation. Self-location and perspective. Body ownership and agency. Free will. The self as a good regulator. The self's blankie. Subjectivity of self-models. Phenomenal content and phenomenal affect. Virtual reality, real suffering. The third revolution.

Notes 77

References 84

4 Speech and Sign 93

The dawning of magic. A shared toolbox. Writing. Stories and culture. Narrative selves. Power, persuasion, and propaganda. Manifesto destiny. The fourth revolution.

Notes 108

References 117

5 Self and Society **127**

A beast or a god. The discovery of critical consciousness. Learning freedom. Selves and collectives. Morality and cooperation. Conscience and consciousness. The fifth revolution. The counter-revolution, or: What Went Wrong.

Notes 144

References 153

Part II The Roads to Freedom **161**

Interlude **163**

At a crossroads.

Reference 164

6 Self-care **165**

Suffering: eliminable and not. Too much self, too little social. The dance of life. The heart of a heartless world. Being and nothingness and zombies. The science of getting help. The sixth revolution.

Notes 173

References 177

7 A World to Win **183**

The status quo. False consciousness and alienation. Critical consciousness: coming to senses. Optimism of the will. Selves in society. Armed love. The last revolution.

Notes 196

References 206

Epilogue **217**

The way out.

Index **221**

Part I

The Human Condition

Prelude

The story of Oh: a brief account of consciousness, from Amoeba to Zoloft.

> Mymosh, thus booted, went flying into the nearby puddle, where his chlorides and iodides mingled with the water, and electrolyte seeped into his head and, bubbling, set up a current there, which traveled around and about, till Mymosh sat up in the mud and thought the following thought: — Apparently, I am!
>
> *Mymosh the Selfbegotten* (from *The Cyberiad*)
> — STANISŁAW LEM

In the beginning there wasn't anything of notice. This was because for things to be noticed, they need to be told apart from other things, but telling-apart hadn't been invented yet.

Every now and then, a tentative telling-apart would rise above the background hum of mindless goings-on, only to disappear back into it without a trace. A telling-apart played itself out every time a pebble, dislodged by the wind from its perch above a boulder on a hillside, would take *this* as opposed to *that* route around the boulder. But such one-off incidents were of no more lasting significance than patterns of cracks in drying mud are for the next day's rain. And none of the parties involved — the pebble, the wind, the boulder, the hillside — cared: caring hadn't been invented yet, either.

It went on like that for a long time.

The first real telling-apart happened when a random patch of slime, just two molecules thick and bobbing in a puddle of vaguely organic matter, closed in upon itself. As its inside became definitely distinct from the outside, a membrane-enclosed cell was born. Finally, there was an event of notice — first and foremost to the first-ever noticer, the cell itself. The cell couldn't help noticing its coming into existence because its very life was a process of telling-apart: the discernment between *itself* and the rest of the world. In that, it needed no external validation: the closing off of its membrane was all that it took.

From the capacity to tell things apart in space, it was only a short step towards the capacity to tell them apart also in time. Protected from the environment, yet responsive to it, the cell's innards retained impressions for a while, making it possible for the cell to compare what it sensed to what it remembered, and to discern things present from things past.

Along with discernment and memory, feelings and caring came into being. An amoeba that preferred certain things over others, such as staying alive

S. Edelman, *The Consciousness Revolutions*,
https://doi.org/10.1007/978-3-031-24012-6

over dying, ended up making more copies of itself — preferences and all — than one that just didn't care. A real preference, though, is one that is backed up by an urge to do something about it and a willingness to do it; and caring is only real to the extent that the carer puts its metaphorical money where its equivalent of a mouth is. When amoebas spend their energy to stay alive and to cater to their other preferences, this is precisely what they do.

Sentience — consciousness at its most basic — is no more than having feelings and caring about certain distinctions, which one makes on one's own. This is why single-cell lifeforms have been basically conscious from the get-go. It did not stop there: built-in capacity for discernment, when validated by caring and backed by memory, is potent magic. Consciousness turned out to be so good at protecting its wielder from foes and from the elements that the evolutionary arms race kept fielding ever fancier versions of it. (It still does.)

A major advance on that front was the invention by single-cell critters of virtual reality. The seat of an amoeba's speck of sentience, its command-and-control machinery, is inside its membrane, but much of the action on which its survival depends happens outside of it. How can it get vital intelligence through and make the most of it? By building on the inside a modest virtual-reality model that tracks the neighborhood presence of food, pheromones, and whatever else in its little world that it cares about.

Now, evolution has always favored sentient beings that care the most about their own welfare (and, depending on the species, perhaps also about the welfare of a few deserving relatives and neighbors). This is what pushed the model of the world to include — as its most important part, no less — a newfangled addition: a model of one's own body. But a model of the body is of little use if it does not include also a model of the body's command-and-control system, which in turn contains a model of the world, inside which there is a model of the body, which in turn contains a model of the world, inside which there is a model of the body, which in turn contains a model of the world, inside which there is a model of the body . . .

At this point, evolution felt dizzy and had to sit down.

In one key respect it did well: by taking the modeling of the world seriously, the virtual-reality *self* that it just invented made conscious control more effective. There remained one little problem, the infinite regress: like any exercise in self-reference, a self is a snake that tries to swallow its own tail. A self that plans its host's conduct by predicting how the world will act in response to it would have to predict its own steps first. An impasse!

A simple remedy for that was discovered soon enough: keep the actual planning shallow, while wiring the self so that it *feels* it's in full control. Unfortunately, this also makes the self automatically the bearer of all responsibility for everything that comes out of its host's actions — a side effect of being self-aware that really helps learning, at the cost of condemning a conscious being to anxiety and suffering.

The path to the next stage of self-awareness opened up with the invention of language. As it always happens with revolutionary technologies, the fallout from the side effects of having language at one's disposal rivaled the gains made possible by its primary use. Using language, the conscious self could spin stories about its life and times, boosting its effectiveness as the ghost in control of the machine, in part by subjecting itself to detailed, eloquent, and protracted self-recrimination.

Language also made it possible for many of its users to band together in trying to understand their world, with an eye towards controlling as much of it as possible — including each other. Some of the stories that grew out of that game were about how its players came to be what they are, from the roots of telling-apart to the invention of storytelling and of its perennial use in the pursuit of power and profit. Eventually, as consciousness itself and the uses of it were taken up by some particularly reckless story-spinners, everyone — both the users and the ones being used — found themselves facing a mirror, and in the mirror a revelation: "Oh! So *this* is what it's all about!"

❧ ❧ ❧

1 Selfless Consciousness

Tower of power. The brain's virtual reality. Identity theory. Computation all the way down. No cognition without representation. What makes representation special and how it can give rise to basic phenomenal awareness. Stimulus and response and awareness and zombies. Slipping into the future. Once again, with feeling. The Gang of Four. Minimal consciousness. The first revolution.

> About him rose the Pastel Towers, tall and gracefully shaped to mathematical curves, tinted pale blue or fuchsia or dove-grey. They reached up for hundreds of feet, cut with quaint and complex designs that some said were the high point of an inimitable art, thought by others to be representations of the actual geometrics of Time.
>
> *Viriconium*
> — M. JOHN HARRISON

Tower of Power

CONSCIOUSNESS is complicated. To understand what it is and how it works, we must first sink a shaft all the way down to the bedrock, to the mindless matter from which it arises. We can then carefully work our way back up, from the foundations to the pinnacle, noting how each new layer of complexity adds sophistication to the capacities, and richness to the experience, of being a fully conscious, alert, and focused human.

The observation platform on top of the tower of human consciousness, such as the one your narrator is perched on at this moment, commands a broad view. In every direction, a dark and fantastic urban landscape, which looks like a collaboration between Piranesi and Dalí, stretches all the way to the hills that rim the horizon. From the maze-like casbah down below, teeming with obscure activity, there emerge tensile, tapering, telescopically jointed towers of attention, which grow tentatively, only to retract back into themselves after a while. As a storm of self-awareness gathers, the spires sprout little tongues of St. Elmo's fire, whose uncertain light is drowned at intervals by bolts of lightning spanning the tower tops and the low clouds.

This city of consciousness, the ultimate soft machine, is such a wonderful mess because its architect — evolution (motto: "Whatever works") — respects no zoning laws or safety regulations and follows no master plan. There

S. Edelman, *The Consciousness Revolutions*,
https://doi.org/10.1007/978-3-031-24012-6_1

is, however, a common pattern to all instances of naturally evolved consciousness. Any living entity is better off sensing its surroundings, having feelings about how it goes (some good, others not so much), and caring enough to act on them than being unheeding, unfeeling, and uncaring. Basic *phenomenal awareness* of the environment, which consists of sensing, feeling, and caring, is the cornerstone of all natural conscious experience, including ours.

As a human, I spend at least part of my waking life being *self*-aware, thinking explicitly about myself, my experiences, my plans, my actions, and their repercussions for my future. This kind of person-level conscious experience does not, however, encompass everything that's going on in my mind, not by far. In the expansive casbah at the foot of the self's tower, many feelings, thoughts, and even social interactions unfold without the involvement of the self. The ghostly crew that makes this happen may haunt person-level consciousness, but it rarely intrudes upon it. Subconscious processes — *subpersonal* would be a better term — are perfectly capable of minding the body's routine business and of monitoring the surroundings for any anomalies that might necessitate an appeal to the conscious self in the tower of power.[1]

Nor is the very existence of the tower of self — that is, of person-level consciousness — continuous in time. There are occasions when the conscious self's purview dwindles to nothing and it is let go of. The cloud-capped tower vanishes entirely, leaving behind a wrack of raw feelings and cares, such as are also experienced by all other sentient beings — even scallops or amoebae, which sense the world and have a stake in the outcomes of their actions, yet are never *self*-conscious like we are.

Amoebas, just like humans, can be more alert or less so and can even be put under an equivalent of general anesthesia, and by the same chemical means, too: volatile compounds such as chloroform.[2] Likewise, an alert scallop in its natural habitat may feel — be aware of — the movement of the current, the changes in salinity caused by influx of fresh water from a nearby estuary, and perhaps the shadow of a passing cuttlefish; an anesthetized scallop will feel nothing. When I am in a state of selfless consciousness, then, my experience becomes akin to that of an edgy scallop, having a care about possible predators in the water above it, or an alert amoeba, feeling drawn in the direction of potential dinner.[3]

It may be that with proper training, I would be able to shed even more of the trappings of regular animal wakefulness. Letting go of the feelings and the care evoked by the sensory world would leave in place a minimum of consciousness: the mere registering of the sensorium; or perhaps even less than that — just pure, content-free alertness.[4] And when that too vanishes, as in deep sleep or anesthesia or death, no consciousness at all is left. The bedrock has been reached.

The Brain's Virtual Reality

To be a rock and not to roll — what is it like? For a conscious being, empathizing with boulders is not easy. What it feels like to be a rock is something we can reason about, but it is pretty damn hard to imagine, and utterly impossible to experience. This is because for all sentient beings conscious experience *is* existence, and feelings are what experience consists of. No feelings, no existence. The moment you truly succeed to feel what the rock feels, which is precisely nothing,[5] you cease to exist. If you decide to try this at home, make sure you have a buddy standing by to resuscitate or reboot you, and hope for the best.[6]

Imagining not to have any feelings is imagining out of existence not only the sum total of one's own being, but also, as it happens, the entire universe. For as long as I am conscious, the universe exists *for me* to perceive and act on; and when I cease to be conscious and descend into oblivion, whether temporary or permanent, everything disappears with me. My conscious experience, then, is not so much something that I have, or even something that I am: it's what everything is — for me.

How does that everything — one's subjective reality — relate to the outside world that most people believe exists independently of anyone's state of consciousness? The common and intuitive assumption of the existence of an observer-independent real world out there includes an equally intuitive rider: that it is being watched as if through a window or a video monitor (to pick vision as an example), so that everything that is going on outside registers directly and faithfully on the inside.

Come to think of it: when my eyes are open, it does seem like I am looking at the world through a wide opening in the front of my skull, right where my nose is, as verified by independent means.[7] As I most definitely do *not* have a hole in my skull, the assumption that sensory experience is as direct and immediate as a look through a window is thereby negated.

The analogy with a video monitor survives this simple exercise in introspection, only to be undone by another one, this time focusing on what happens to my visual world when I open one of my eyes or the other in alternation, while looking at a nearby object such as the back of my hand. What I see is this: against the background of the room, my hand seems to shift left or right, depending on which eye I am looking through at the moment. Clearly, the views through the two eyes differ. This is as it should be: the eyes, after all, differ in their position on my face.

Returning now to the view that I get with both eyes open, I realize that it is not the same as either of the two monocular views. There is, for one thing, *more* to it: with two eyes, I can see depth! Depth information is, of course, available implicitly in a pair of monocular views in the form of differences in the positions of a given feature (such as my finger) in the two views, called binocular disparity. The process of binocular stereopsis, which combines two disparate views to *infer* depth, uses disparity to make depth information explicit, through a series of computations that are by now well-understood.

The video monitor analogy no longer seems like a good account of what is going on here. The part about conveying a signal from where the camera is to another place still applies. Phenomenally, that other place is my perceived self-location: inside the skull, a few centimeters behind the invisible to me bridge of my nose. (That, incidentally, is also where my physical brain resides, without whose activity there would be no phenomenal awareness and no conscious self.) The part about inferring information from the signal does not, however, fit the notion of mere monitoring. What do we call a self-contained visual world, to which information that is not readily (or not at all) available on the outside is added, and in which a viewer is immersed? A virtual reality.

Identity Theory

Having remarked just now so nonchalantly that without brain activity there would be no consciousness, I better go ahead and clarify my stance on what the public perceives to be a big question (perhaps the biggest one) with which the science and philosophy of mind contend: the mind-body problem. The complete dependence of the mind on the brain seems obvious: even those who profess a belief in an immaterial and immortal soul are typically extremely reluctant to have it tested ahead of their allotted time. But what does this dependence consist in?

Because an anesthetized brain (let alone a dead one) is quite devoid of any mind-like quality, it is clear that physiologically normal functioning on part of the brain is necessary for the mind to arise. At the same time, there does not seem to be anything else that's needed for that to happen, over and above the brain doing what it normally does. In other words, the brain's doing what it does is both necessary and sufficient for the mind to happen. This realization is a sign that we can afford to stop beating around the bush. No more circumlocutions of the kind "the mind *arises* from what the brain does": say instead "the mind *is* what the brain does."[8]

What might that be? We know what all the other organs do: the stomach participates in digestion, the kidneys filter the blood, the red blood cells carry oxygen, etc. What does the brain do? The answer to this "merely" biological question sounds, upon first encounter, fairly at odds with what biology is thought to be about. Here it is: what the brain does is compute; and what it computes is the mind.

The MIND is what the brain does. What the brain does is COMPUTE the mind.

This fundamental insight immediately helps one understand how the virtual reality that the conscious self is immersed in comes to be. Although I still have to explain what is meant by computing (which I shall do in the remainder of this chapter), intuitively it is clear that inference and other kinds of information processing involved in setting up a virtual reality call for computation. In chapters 2 and 3, I shall describe how the conscious self too is a virtual computational construct.

The definitive resolution of the mind-body problem, then, consists in the recognition that the mind is *identical* to what the embodied brain does. When applied to the understanding of consciousness, this resolution calls for identifying each subjective phenomenon with some physical process that instantiates a particular class of computations (which, of course, can be observed and measured from the outside). Thus, at all times, each facet of my conscious experience must be equated with certain aspects of the events in my brain, such that neither the experience nor these aspects could possibly take place in the absence of each other.[9]

It would be nice if such claims of identity turn out ultimately ("when all is said and done") to be convincing, but we should not rely too heavily on intuition in judging the plausibility of scientific explanations. How plausible is it that the same invisible goings-on are behind both lightning and a lodestone's attraction to iron? Not very, unless one has learned about Maxwell's theory of electromagnetism and its experimental corroboration. Likewise, if your common sense revolts against the notion that feelings — *feelings!* — could be explained by the brain's doings, quarantine your intuition at least until the scientific case for the proposed hypothesis has been made and evaluated. And if you counter that explaining the mental in physical terms is not the same as reducing two physical phenomena to a common denominator — that's just another intuition, to be as distrusted as the others.

As we shall see next, computation is the common denominator between the components of consciousness — phenomenal awareness, the feelings that comprise it, self-awareness — and physical processes that implement them. Like all the other aspects of the mind, consciousness is something the embodied brain does; and what the brain does is *compute* the mind.

Computation all the Way Down

Computation is ubiquitous in the universe. It is how the universe figures out how to conduct itself over time, its rules of conduct being the laws of physics, as expressed in the language of mathematics. Because computation is so fundamental and indispensable to the understanding of any and all aspects of the world we inhabit, we need a simple and effective way to think about it, which would apply across the board and not just to computers or, for that matter, brains. Here's one: computation is a sequence of constrained *transitions* between the *states* of the system that carries it out.

Instead of formally defining the two key terms that I just invoked, transitions and states, let me offer some examples and let your imagination fill in the details.[10] Think first of the abacus, a mechanical device for doing arithmetics — a wooden frame with several strings threaded with some wooden beads strung across it. On each string, the beads can be pushed all the way to the left, or all the way to the right, or they can be split between the left and right sides. The arrangement of beads — this many on the left, that many on the right — is the *state* of the string they're on; the state of the entire abacus is the combined states of all its strings. Computation then corresponds to making *transitions* between states by moving beads around to implement addition or subtraction, the constraints being the rules of arithmetics.[11]

An electronic digital computer works much the same way, except that instead of beads and strings it has an array of memory cells. As before, computation is a series of transitions between states — changes in the pattern of zeroes and ones held in memory. Unlike an abacus, though, an electronic computer does not need outside help to make the transitions happen; instead, these are determined (constrained) and carried out by the machine's circuitry on its own. This capacity of certain computational devices for autonomy bodes well for the prospects of applying the concept of computation to the understanding of how the mind works.

No less important is the possibility of the states of a computing system being graded rather than discrete (which is what they are in an abacus or a digital computer). For example, the state of a mechanical system consisting of a falling apple and the earth is naturally captured by position and momentum variables, which are "real-valued" (that is, generally require a long sequence of zeroes and ones to specify to the requisite precision). The state of a brain can likewise be described by a (very) long array of such numbers, each corresponding to some electrochemical dimension of the physiological state of affairs in one of the brain's cells.

The possibility of encoding a state by an array of numbers suggests an immediate and extremely effective way of visualizing the inner life of a computational system. It's a quite straightforward exercise in geometry: just like two numbers can be interpreted as designating a location on a map if treated as longitude and latitude values, a list of seventeen numbers stands for a location in an abstract seventeen-dimensional space. For instance, if you have rated seventeen movies on a scale of one to ten, your ratings jointly code for a point — your location in a seventeen-dimensional movie ratings space. Getting a thousand people to do the same yields a cloud of a thousand points in that space (the structure of which can then be used to generate movie recommendations or even to assist matchmaking).

Now, to conclude the exercise, consider this: if it takes a million numbers to describe the physiological state of the brain, it can be thought of as a point in a million-dimensional abstract space (I know, this has escalated quickly). That point is just one momentary state, frozen in time. Unfreeze time, and the brain will compute its next state from its past history (including any interactions with the world). The process of computation in the brain can then be described in its entirety by a *trajectory* through the brain's rather large state space.

No Cognition Without Representation

The computations performed by the brain are mostly quite boring, having more to do with routine maintenance than with fancy stuff like consciousness. To zero in on the latter, we must start with what I call *cognitive* or mind-like computations.

Cognitive computations, about which I have written at length elsewhere,[12] are easy to recognize as such. A computation qualifies as cognitive if it involves *representations*: symbols or patterns that stand for something else than themselves, by virtue of being interpreted as such — that is, made meaningful — by the system that maintains them. And what should a cognitive system be representing? Why, the world, of course! Or at least as much of the world as it needs and can afford to represent to live long and prosper.

For the human brain, the need to represent the world by some internal means is obvious. The brain is shut inside the dark confines of the skull and must therefore base its conduct on the measurements provided by the senses.[13] But my favorite example of a cognitive system that represents its world[14] internally is much simpler and would surprise those who think of cognition as something that only humans do. It is the *E. coli* bacterium. In this little animal, outside temperature and oxygen level are represented by molecular interactions inside the cell. These biochemical processes modulate gene expression and do it on the fly, allowing the bacterium to fine-tune its metabolism and behavior to the prevailing environmental conditions.[15] Not much of a mind, perhaps, but all of it is the microbe's own.

What Makes Representation Special and How it can Give Rise to Basic Phenomenal Awareness

Like Pavlov's dog, the *E. coli* bacterium not only responds to stimuli presented to it by the environment, but can also learn to associate familiar stimuli with novel responses.[16] More than that: its internal state transitions track outside changes as they happen in real time. This ability is crucially important for the *E. coli*, whose typical life cycle involves abrupt shifts between a warm anaerobic environment in its host's gut and a relatively cool, high-oxygen one out in the open.

The concept of representing the world internally is the key to understanding the hard core of consciousness: phenomenal awareness. For me to be aware of a situation must mean to be in an internal state that represents it. If the state in question is not internal to me, it cannot be *me* who is aware; and if the state does not represent something, it cannot be an awareness *of* anything.

Reducing phenomenal awareness to representation would trivialize consciousness if cognitive representation were trivially easy to achieve — but it ain't. Does the appearance of a "slow down, sharp curves ahead" traffic sign in the margin of this page mean that the book you're reading is aware of some road hazard, or of the sign's metaphorical reference, the conceptual complexity of the flagged passage? Of course not: the requisite relationship between the state of the page and the world is established not by the book but by you, the reader. The same goes for the state of the memory chip in your phone where your selfie is stored. The numbers in memory, which code for pixel values, represent a face for you (and then only if displayed on the phone's screen), but not necessarily for the phone itself, which could be programmed to interpret these numbers in an entirely different manner: not as an image, but as the password to the phone, or to a bank account.

The lesson here is that representation is only "for real" if the system's internal states are interpreted by the system itself, as for instance when your brain state that ensues when you look at the "slow down" sign causes you to slow down (while driving, or while reading a convoluted passage). This requirement — that the interpretation or *meaning* of a representational state be internal or *intrinsic* to the system — rules out scenarios that would otherwise drain the concept of representation of all explanatory power. It also bodes well for the project of applying the concept of representation to the understanding of phenomenal awareness. Not being up to an external observer is the single most important requirement of conscious states. However *you* may choose to interpret any of *my* phenomenal states, what they really mean is what they mean to me. Likewise, I have no say in deciding what it is that you are really conscious of right now, even as I am trying to affect your brain state from the outside, through these pages.

This is not simply a matter of an outsider having sufficient physical access to your brain: there is a genuine computational distinction between what it feels like to be you *to* you and what can be concluded about it from the outside. This distinction, which underlies the sense of privileged access that individuals have to their own phenomenal states, arises from the basic need to set the system in question — an embodied brain, or a microbe — apart from the rest of the world. Computationally, this setting apart results in a partition of the set of variables that completely and exhaustively characterize the entire physical universe into two: those that are internal to the system, and the rest, which are external to it.

"Before we can talk about anything, we have to consider what distinguishes a 'thing' from everything else."

— Friston et al. (2020)

From the outside, only a subset of a system's internal variables would typically be accessible to influence; these are the sensory variables. Conversely, a system would only be able to directly influence a subset of the rest of the world's variables; these can be called the active variables. Together, the sensory and the active variables form a so-called Markov blanket that interposes between the system and the world.[17] In the case of an *E. coli*, the blanket consists of the sensory and active variables implemented by the membrane, which physically encloses the cell's interior. In the case of a human brain, the sensory variables correspond to the photosensitive cells in the retina, the hair cells in the inner ear, and other world-to-brain information transducers; the active variables are implemented by the muscles.[18]

Now, the specific causal connectivity of the Markov blanket — indeed, its very existence — constrains the probabilistic transitions among the states of the internal variables in such a manner that the system's access to its own internal states becomes privileged in a concrete mathematical sense. This, then, is the first of the two distinguishing features of representation, which make it both non-trivial and uniquely capable of supporting phenomenal consciousness.

Having dismissed the notion that arbitrary attribution of meanings to system states by an outside observer counts for anything, we can take on another commonly held misconception: that a mere momentary co-occurrence between a system's state and a state of the world is a sufficient basis for a representational relationship. Not only are momentary isolated co-occurrences meaningless because omnipresent (after all, at any instant, the system is in *some* state, as is the world) — this situation does not allow for any interpretation, precisely because it is momentary and singular, and interpretation is a process that must involve at least two moments in time. In a true case of representation, a system state that has been brought about by a world event (or by another internal state) consistently leads, in time, to a succession of states that constitute its intrinsic meaning — the meaning of the event for the system. The requirement of consistency over multiple occurrences of the triggering event rules out as meaningless mere one-off coincidences between system and world states, which signify nothing.

This mathematical requirement highlights the other sense in which representation is special — and specially suited for supporting conscious awareness, on part of the system, of changes in the outside world. Representation is special in that it necessarily involves not isolated system states, but rather sequences of states — in other words, trajectories through the system's state space, a concept that came up on p.12 above, in the context of defining computation. It is important to note that a consistent, recurring pattern of state transitions is extremely unlikely to arise as a result of chance (the more so, the larger the state space and the longer the pattern).[19]

Every instance of representation that is not a statistical fluke must therefore be backed by some causal mechanism that puts the pattern in place. In natural representational systems, from amoeba to zebra, this mechanism takes the form of cascades of biomolecular events, which implement the relevant signaling pathways. These, in turn, make it possible for the animal to learn the reliable representations it needs, from experience. In each such case, the ultimate guarantor of proper causality, and of the reliability of the learned representations of events in the world, is evolution.[20]

Evolution is also the force that drives the emergence in natural cognitive systems of the capacity for *discernment* — telling some world states apart from others, as needed to promote survival and maintain fitness. Is this berry ripe or unripe? Is that a leopard peeking through the dry grass or just some clumps of dirt? To answer such questions, which may be of vital importance, the perceiver must compute representations that are not only intrinsically meaningful and reliable, but also intrinsically indicative about the corresponding differences among certain world states: ripe/unripe, leopard/dirt.

We can now connect the dots leading from the concept of representation to that of consciousness. To do so, observe that any system that consistently and reliably registers, and intrinsically interprets, differences or changes in the outside world can hardly be described as unheeding of what is happening around it. It makes sense, therefore, to call such systems *minimally phenomenally aware*. Minimally, because if any of the components of this level of awareness are missing, there goes the awareness. Phenomenally, because, as I shall argue next, nothing else is needed to account for what it *feels like* to be aware.

Any system that consistently and reliably registers, and intrinsically interprets, differences or changes in the outside world can hardly be described as unheeding of what is happening around it is MINIMALLY PHENOMENALLY AWARE.

Stimulus and Response and Awareness and Zombies

How can you tell whether or not I am truly phenomenally aware of the ripe berries I am looking at — that to me right now they *feel* good-to-eat? Perhaps I am merely going through the motions: calling the berries ripe, and thus making an outward discernment, but without actually experiencing anything on the inside. This quandary, which is an aspect of the so-called problem of other minds (a relative of the mind-body problem in philosophy), refuses to go away in part because overt behavior is indeed not a reliable indicator of any internal representational processes — let alone consciousness, for which the explanatory bar is so much higher.

A vivid illustration of the difficulty we are facing here is the conceit of philosophical zombie (which has only one thing in common with Hollywood's brains-craving undead: both species are extremely difficult to dispose of). A p-zombie is an imaginary creature that is supposed to be just like a normal human being in all physical respects and is also mentally on par with humans, with a single exception: it has no feelings — no phenomenal awareness — and is incapable of any kind of conscious experience. Quite a few philosophers of mind believe that just because p-zombies are conceivable, the concept merits serious discussion.[21] Those who accept this claim believe they are owed an explanation of what it is, over and above the normal operation of the brain, that bestows consciousness on a creature that would otherwise be devoid of any phenomenal inner life.

To appreciate the flavor of the controversy surrounding p-zombies, I suggest that you imagine a related species, the dance zombie. Unlike real dancers, from whom they are physically indistinguishable, d-zombies do not really dance even when they do all the right steps at the right time — rather, they merely go through the motions. By analogy with the p-zombie argument against explaining consciousness in terms of what the brain does, one's ability to conceive of a d-zombie would imply that to be real, dance must possess some special, presumably non-physical, ingredient, over and above mere patterned motion on the part of the dancer.

Of course, there isn't, nor there needs to be, any mysterious extra ingredient that turns "mere movement" into dance. In this respect, being phenomenally aware is like dancing: going through the motions is all it takes. What motions would those be? Those of computation in any system which, like the brain, consistently engages in intrinsically meaningful and discerning representation of the world by internal states. The identity theory holds that registering a stimulus and discerning it from other possible stimuli by traversing a representation space in response to it ("going through the motions") in a specific manner is *identical* to being phenomenally aware of the stimulus. In particular, for two stimuli to be experienced as distinct, the representation-space trajectory segments they evoke must differ intrinsically and qualitatively.

A stimulus that gets registered in this manner makes a difference where it matters — to the system itself — by having an effect on the unfolding of the system's internal state-space trajectory. This type of internal "response" (which may or may not lead also to an overt one) has a special significance in so-called *dynamical systems*, which by definition compute their next state, given their history (that is, the past state-space trajectory) and the external influences (if any). Examples of dynamical systems include the apple-and-earth duo ensemble, as well as the hugely more complex *E. coli* bacterium, but not the abacus, which would compute nothing if it were not for its beads being watched and shifted around by an outside agent.

In the human brain, the vast state space, which is spanned by the activity of its billions of cells, has room for trajectories of unimaginable complexity. Nevertheless, there is a certain simple pattern to the temporal dynamics of phenomenal awareness generated by the brain. The pace of human consciousness over time has been compared to a bird's life, which alternates between "flights and perchings." As brain cells interact with each other and sometimes with the outside world, one experiential state ("perching") follows another, each lasting for much longer than the transition ("flight") to the next state. The perchings, of which there are several every second, allow the brain the time that is needed to process its present state.[22]

It is worth stressing again that in the brain as in other dynamical systems, the procession of states is being monitored and acted on at any time by the system itself. A neuroscientist can try to guess what is going on by monitoring state change events from the outside and coming up with interpretations for them, but what the brain's constantly shifting patterns of activity really mean is up to the brain itself. A particular brain event that has no consequences for other brain events down the line is truly of no consequence — it might as well have never happened. If a chain of events does develop, it may or may not contribute to conscious experience, but in any case, it's entirely beside the point how an outside observer interprets the corresponding brain states: how my brain interprets its own states is the only thing that matters *for me* — that is, for my thoughts and my feelings.[23] The self-interpretation property explains why even the most primitive kind of consciousness — the non-egoic minimal phenomenal awareness of the world, from which the self as such is as yet absent, is already self-directed and self-sufficient.

Slipping into the Future

The ability to represent not just differences between objects or events but also the dynamics of the environment by means of corresponding internal dynamics is much more difficult to achieve (and infinitely less likely to arise by chance) than taking static snapshots of outside conditions. This makes the capacity for phenomenal awareness, which I equate with intrinsic, dynamic discernment, so rare in nature, our predisposition to animism notwithstanding. Rarer still than basic phenomenal awareness are higher-order levels of consciousness, such as the awareness of awareness, or the awareness of the perceiving, thinking, and acting self (see Figure 1.1). Most of the computational processing that constitutes the human mind is unconscious in this higher-order sense, but only in a coma are we ever completely unaware.

There is a good evolutionary reason behind the impossibility of all of the mind being unconscious all the time, as supposed to be the case for a p-zombie. Dynamic representations, which are the core of phenomenal awareness, are uniquely suited for dealing with environmental challenges, because they can be used to *predict* the future. This ability taps into a very basic property of this universe, which has a degree of predictability built right into its foundations in the form of the countless natural processes whose dynamics is patterned rather than random. Ripe apples fall to earth, night follows day, prey comes to drink at the water hole, predators like to be downwind from you... there's no end to such patterns, many of which can be exploited. In such a setting, past performance of a behavioral tactic is sufficiently predictive of future returns to make betting on success worthwhile. All that a creature needs to do to practice foresight, thereby gaining advantage over competitors that lack it, is to represent internally the dynamics of those of the chains of outside events that matter to it the most.[24] And that, in turn, makes it at least minimally phenomenally aware.

Once an internal representation of some patterned natural process is in place, it can become active even in the absence of external stimuli. As time rolls, the dynamic representation — an internal simulation or *model* of the process in question — will churn out predicted states corresponding to world states from farther and farther on in the future.[25] To be sure, the more remote the future, the more off-target the predictions are likely to be. Still, reliable guessing ahead even for a fraction of a second could be helpful for creatures whose life is lived in the fast lane.

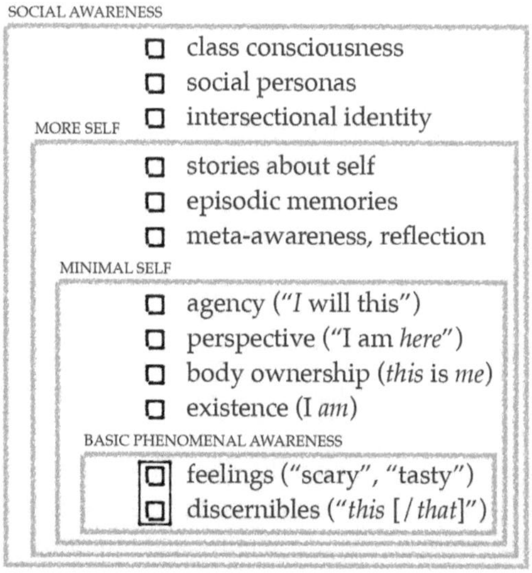

Figure 1.1 — Levels of consciousness, from basic phenomenal awareness (this chapter), through the minimal self and full-blown, language-capable and self-aware one (chapters 2 through 4), to socially embedded selves and class consciousness (chapters 5 and 7).

Once Again, with Feeling

Because foresight would be of no use at all to its practitioner if it were not tightly coupled to action, evolution ensures that it is. As we shall see, the principle that governs this coupling is directly responsible for the single most insidious trick that evolution ever pulled on its creations. If there is any way for us to make up for this original sin, of which we are not perpetrators but victims, it can only be discovered by scrutinizing the relationship between foresight and action.

Intuitively, having foresight promotes survival and flourishing by making it possible to tell apart actions with good projected outcomes from those with bad ones. But what does good and bad mean in the first place? Assuming that the basic ability to discern *value* is innate merely kicks the can down the road. Still, it is evolution that is ultimately responsible for the emergence of value. The knowledge of good and evil must have originated in the plain old Darwinian "selection by death": missteps punished by throwing the actor out of the genetic pool.[26]

Under such a harsh regime, having some notion ahead of time, in a way of foresight, about the eventual value of an action — as in "How likely is it to kill me?" — is a step in the right direction. There is, however, one essential component that is still missing from this purely informational version of value: *caring* about the information it contains. In systems that evolve under selection pressure, caring takes the form of a *drive* to avoid inauspicious (low- or negative-value) outcomes.

If value is to be effective rather than merely informative, this drive must be made difficult to resist. It cannot, however, be unconditionally irresistible — that would make the resulting behavior too rigidly determined by the evaluation mechanisms. Imagine yourself catching a glimpse of some ripe blackberries in the midst of a thorny bush. Getting through the brambles to the berries would be painful, but forgoing an attempt to do so would be costly in another way: you are famished. What should you do? If your brain's action planning system is confident that its estimate of the berries' value makes them worth the price, it may decide to tolerate the pain of pushing through the brambles, as it spikes and ebbs, until the reward is within reach.

The prime challenge in this situation is balancing the pain against the gain. The confidence of the action planning system is a key factor in seeking the proper balance: if the confidence is high enough, pain should be temporarily ignored. This creates a problem: how to ensure that hunger — signaled by a glucose-monitoring mechanism that is oblivious to all other aspects of the bodily state, including pain — does not always win. A forager that fails to address this problem will sooner or later bleed to death, impaled on the thorns of a bramble thicket that happened to have exacted too high a price for the glucose boost it provided.

The problem at hand is essentially about trust in signaling: in this case, the decision maker's need to determine the honesty of the "daring" faction in communicating its confidence in the plan to push on despite the pain. Honest signaling happens to be a very well-known problem in evolutionary science. It arises, for instance, in the context of a typical courtship ritual: how can the female trust the male's display that is intended to convince her of his superior quality? According to the Handicap Principle, proposed by Amotz Zahavi in the 1970s, the signaler can be trusted if the signal is such that it would be costly to fake — as in the case of the peacock's tail, which requires considerable resources to grow and places a literal handicap on his ability to evade predation once fully grown.[27]

The Handicap Principle has been invoked as a possible account of the evolutionary origins of pain. Because estimated value is a kind of signal, used by certain brain processes to influence decision making, its reliability needs to be ascertained; the same goes for the quantity to which it must be compared, namely, the expected tissue damage, which is signaled by another set of processes. A handicap can be placed on the signalers by requiring that they bid on participation in decision making — that is, pay up front a cost, which "low-quality" signalers cannot afford. On this account, then, pain *is* the phenomenal experience of this cost.[28]

Pain: *per aspera ad astra*, or: the cost of confidence.

The Gang of Four

For this latest appeal to identity theory to work, it should include an explicit account of the computational underpinnings of value, caring, and feelings in general. Given that feelings always accompany any action, be it overt (behavior) or covert (thinking), a proper computational account will necessarily encompass all of these. There are four functional components to this account. The **first**, foundational one is the internal representation of the environment, which needs to include a record of previous engagements with the world — a kind of *memory*. A dynamic process of modeling the world turns memories of the past into predictions about the future, which in turn support foresight. Because the representation space trajectories can track the dynamics of various developments of concern, they can be used to predict where things are going, so to speak, simply by extrapolating the present tendency into the near future.

The predictions generated by this dynamic memory are the basis for deciding what to do next. Each alternative course of action is championed by an instance of the **second** functional component, an *actor* process. It is these actors that vie for access to the actual behavioral control mechanism, by bidding in proportion to their confidence in the value of the expected outcomes of their respective actions. This value can be estimated by keeping a running ledger — another form of memory — of associations between states and actions on the one hand and, on the other hand, the outcomes they have led to in the past. By default, this ledger incorporates certain preconceptions about how the world works, built right into the system's basic anatomy and physiology. These innate biases towards a certain class of solutions to behavioral problems capture the relevant qualities of the ecological niche in which the species has been evolving. They anchor the lifelong process of learning to attach value to experiences.

It is up to the **third** functional component, the *decision maker*, to select the winning actor or actors, which are then tasked with supervising the behavior aimed at the chosen goal, while taking into consideration the relevant resources and constraints (the body, along with knowledge about what it can and cannot do in the given situation). Planning and execution are typically carried out concurrently. If the most recent projection of the state-space trajectory is estimated to lead to no good, a corrective action is sought. Overt behavior takes place in the real world: in the physical space in which the system lives. But an action also has repercussions in the internal representation space, where it causes a scramble to generate new predictions and compute their consequences.

As the planned behavior takes shape, the **fourth** functional component, a distinct internal process called the *critic*, revisits the choice, the plans, the action execution, and the ensuing value, the latter being determined by the representation space neighborhood in which the system finds itself. In its judgments, the critic too relies on past associations between sensory qualities (red, salty, far-away, loud) attached to states and the outcomes (tasty, exhausting, hurtful) that they evoke. Ongoing updating of this information amounts to learning from the consequences of behavior, or reinforcement learning.[29]

Each occasion on which foresight is exercised is an opportunity for this inner Gang of Four — the memory-based predictor, the gaggle of actors, the decision maker, and the critic — to stage another round of their game. Because evolution has little patience for species in which the critic is at liberty to withhold judgment or to whitewash over failures, this psychodrama often resembles a Greek tragedy: good intentions may end up being punished and initially inscrutable Fate always has the last word. Has the push for berries in spite of the pain paid off? The critic rewards the responsible actor by topping up its reservoir of confidence (as well as updating the value estimate of the reaching action in a generic "berries nearby" situation). Has the pain been too much and the berries not worth it? The critic puts the overly eager actor in its place. The abject feeling of being in the state-space equivalent of the doghouse is inexorable and is all the more poignant for being impossible to dismiss as due to unfair treatment by strangers: after all, the source of the negative emotions is part of the system itself.

Minimal Consciousness

The emergence of emotion-laden phenomenal awareness from insensate and unfeeling matter was the first and most radical of the several revolutions that by and by made human-style consciousness possible. The momentousness of that first awakening suggests a comparison with the mother of all beginnings in biology, the emergence of life. The comparison is apt, but not for the reason most people might expect it to be. Once a mystery, the difference between living and non-living systems is now understood to arise from the organization, or rather self-organization, of certain arrangements of otherwise quite ordinary matter. Likewise, the difference between no consciousness and minimal consciousness is in the kind of self-sustained computation that goes into the latter.

Just like life could not have possibly been explained by an appeal to any magical "vital force" or *élan vital*, consciousness cannot, as a matter of principle, be explained in terms of some secret ingredient that itself already contains it. For an explanation of consciousness to be non-circular, it must describe how it can emerge from what goes on routinely in mindless matter. And because we already know that the rest of the mind is made of computations, we know what kind of goings-on to look for.

The barest minimum of conscious experience — phenomenal awareness that involves feelings and caring — need not include reflection on experience, let alone the erection of a full-blown tower of self (in comparison, advanced consciousness, to be discussed in later chapters, always involves the minimal kind). Even a microbe can have minimal phenomenality, although the range of the aspects of the world that it is aware of and of the emotions that it experiences is rather limited, compared to ours. (I have purposely refrained from using scare quotes when referring to microbe emotions just now. For any of our single-celled gut commensals, feeling so stressed by a situation as to try to swim away from it is feeling enough.) As to humans, the closest we can get to having no more than a minimal experience, short of experimenting with a sensory deprivation tank, is by sitting at ease in a dark and quiet room as a pure tone is sounded. Becoming aware of the tone without reflecting on it (or about oneself, or anything else) is what a minimal experience is like.

Like all types of conscious experience, the minimal one is variable in its composition and graded in its strength. It needs most of its components to be there most of the time, rather than all of the components all the time; and each of those components itself fluctuates on a scale, rather than being all-or-none. These components are all computational, and as such they are functional: that is, they are defined in terms of what they do, not what they are made of.[30] A provisional list of the functions includes dynamic representation, discernment, and caring — the innermost box in Figure 1.1.

MINIMAL CONSCIOUSNESS =
= dynamic representation
+ telling apart
+ caring

Of all the computations that make the world go round, the most important contribution to consciousness comes from the kind I called mind-like — those that deal in representations. As noted earlier, a system's computational state represents some aspect of the external (or perhaps the system's own) state of affairs to the extent that it is interpreted by the system itself. Such self-interpretation does not require the presence or intervention of a self, which is a much more complex computational construct (wait for it until chapter 2). Nothing of the sort: self-interpretation is simply the defining feature of a dynamical system — one that computes and executes its state-space transitions on its own, instead of outsourcing them to an external entity (as an abacus does, to its user). This is the sense in which a computer that plays chess against itself is a dynamical system, whereas a chess board that someone uses to play is not, even though it can be assigned well-defined states and transitions.

Of course, not all dynamical systems have phenomenality, so the mere act of self-interpretation cannot be enough. The other necessary ingredient is suitable meaning. Because the meaning of a system's activity is determined by *how* it is interpreted, there must be something special in the process of interpretation that makes the activity behind specifically conscious processes stand out. My shorthand for this special something is *telling apart*. It is precisely the dimension that any conscious experience must, at a minimum, involve: diagnosis, or the telling-apart of *this* state of affairs from another, or from all others — light from twilight or dark, red from yellow or green, salty from sweet or sour.

Telling apart is something that certain dynamical systems do all the time. Think of your brain, as you sit in the dark waiting for the tone, as being in state A. As the tone sounds, the brain briefly visits another state — say, X if low-pitched, Y if high-pitched — before reverting to A. The difference between the two state space trajectories, AXA and AYA, *is* the experience of the difference in pitch that you, the product of your brain's computational toil, have just undergone.

The human experience of the world is, however, far from being exhausted by a bunch of binary either-this-or-that contrasts. Rather, our experiences reside in a richly structured multidimensional space of graded and interrelated qualities. A key structuring principle imposed on this space by the brain, which maintains it, is similarity between certain points within it (as revealed, for instance, when the pitch of a sound is perceived in relation to a harmonic scale, in addition to its loudness and timbre being registered). Many more qualities spring up when higher-order consciousness kicks in, bringing with it conceptual hierarchies and the semantic finery made possible by language.

Extensive evidence from psychology and brain science suggests that even the most frugal of our phenomenal experiences include qualities that are felt rather than being merely registered or taken notice of. Caring is one common motif to the many dimensions of feelings or emotions.[31] Caring is the fulcrum installed by evolution and used by it to gain leverage over everything that natural representational systems do in a way of controlling their thinking and their overt behavior. What would have otherwise been mere zombie-like information processing ends up being an inner life, always more or less cared about and always felt in various ways.

The First Revolution

The upshot of the computational explanations of representations, of telling apart, and of caring that I have sketched so far in this chapter is that any system whose dynamics fits the bill is capable of sustaining at least a tiny spark of consciousness, of the non-egoic or selfless phenomenal variety. Its emergence signaled the first consciousness revolution. A startling consequence of this discovery is that insofar as basic consciousness goes, being human is a vast overkill, and even being alive in the traditional sense is not necessary. Neither a big brain nor biology as such are required: any dynamical system can do it, as long as its parts interact in just the right way.

This is a rather liberating conclusion, in three respects. First, it disposes of the residual hubris of human exceptionality in yet another corner of social and life sciences. Second, it opens the door to bringing conceptual and mathematical tools from other disciplines to bear on a problem that was hitherto thought to be essentially biological, or perhaps philosophical. Last but not least, it relieves this author from the chore of expounding at length on brain anatomy and physiology, the study of which, as we now know, may supplement, but can never substitute for, computational understanding.

From here, there is still a long way to a full account of the vaunted human consciousness. We already know, however, what that tall tower stands on: its foundation is the minimal phenomenal experience, the capacity for which we share with microbes, mice, and other mindful machines, living or not. The raising of this foundation over the indifferent bedrock is the extraordinary and lasting achievement of the first, and most pivotal, of the consciousness revolutions.

❧ ❧ ❧

Notes

1. The computational processes that comprise the subconscious mind (Kihlstrom, 1995, 2022) should not be confused with Freud's pseudoscientific notions (McCulloch, 1965). As psychological science began to assert its independence from philosophy, the idea that mental processes, including perception and thinking, can be implicit or unconscious (that is, can proceed without the participation of the conscious self) was being promoted by von Helmholtz (1856) (see also Barlow, 1990). Libet (1985) made a notable contribution to the study of decision-making (and to the debate on free will) by demonstrating that a telltale of a "voluntary" movement, in the form of electrical activity in the brain, precedes by hundreds of milliseconds the conscious decision to act, as reported by the subject. (See (Maoz et al., 2019) for a recent critique of those studies.) Lamme (2015) examines the timing of the transition from unconscious to conscious processing in vision. A related and extremely useful theoretical distinction is between subpersonal- and personal-level processes in the mind (Metzinger, 2004). A state of consciousness in which the former dominate is termed non-egoic (e.g., Metzinger, 2021, p.7). What I call, for the lack of a better term, the minimal effective self (chapter 2) is non-egoic in this sense.

2. Cf. Turin, quoted in (Rinaldi, 2014, p.1117): "The only thing we are sure about consciousness is that it is soluble in chloroform." Because ethanol acts on all living cells in a similar manner (Evans and Balster, 1991), you could even get an amoeba drunk. For a comprehensive review of general anesthetics, see (Rudolph and Antkowiak, 2004).

3. Consciousness is sometimes equated with "higher-order" reflection on "first-order" sensory states (Rosenthal, 2009) or with self-consciousness (Friston, 2018). These advanced ways of being conscious can, however, only be understood in the context of the more basic and less human-specific ones, starting with what is sometimes called "phenomenal experience" (Fekete and Edelman, 2011). This refers to the type of consciousness that consists entirely of the awareness of some aspects of the world (and perhaps of the experiencer's body, which of course is part of the world) — a usage similar to that of (Sperry, 1969, 1970). This notion of phenomenal awareness corresponds to what Panksepp (2005, p.31) called "primary-process consciousness." It does not map cleanly on classical philosophical distinctions such as that of Rosenthal (1990) between "creature consciousness," and "state consciousness," or that of Metzinger (2003) between "system consciousness" and "self-consciousness." I will, therefore, refrain from discussing those distinctions here.

4. Pure alertness does not seem to qualify as even minimal consciousness or phenomenal awareness because it is by definition content-free, and as such it cannot be an awareness *of* anything (with the exception of the level of awareness itself). This state, also known as witness consciousness (Albahari, 2009), is prominent in certain meditation traditions. Metzinger (2018; 2020), who refers to it as Minimal Phenomenal Experience (not to be confused with what I call either minimal consciousness or phenomenal experience in this book), offers a philosophical analysis of this state. Alertness is of central interest to Friston (2018), who proposes that all conscious systems continually attempt

to predict their own future level of alertness. This postulate leads him to the conclusion that all consciousness is self-consciousness, which can only stand if both concepts involved, consciousness and the self, are radically reinterpreted. The interpretation I favor is closer to the philosophical consensus, East or West. Thus, for instance, Smart (2004a) distinguishes between basic awareness and "full consciousness" or "awareness (by inner sense) of awareness." In Buddhist philosophy, basic awareness corresponds to what is known in psychology as cognitive unconscious. Thus, Waldron (2002, p.141) links "two core Buddhist concepts, the dependent arising of our 'world of experience,' and the notion of *vijñāna*, 'discerning cognitive awareness' or simply 'consciousness,' ultimately arriving at its most important development in the Yogācāra school of Indian Buddhism, the *ālaya-vijñāna*, a form of subliminal cognitive awareness or 'unconscious [i.e., selfless] structuring of the world.'"

5. There is no denying the difficulty of figuring out what, if anything, a rock feels. This is where a sound theory is indispensable. One family of theories of consciousness, panpsychism, does allow for feelings (or perhaps "proto-feelings") of rocks. On this account, the conscious mind is a fundamental characteristic shared in varying degrees by all matter, down to the elementary particles (see (Goff et al., 2017) for a review of the Western philosophical views on panpsychism and (Gasparri, 2019) for an Advaita Vedānta comparative perspective). To me, venturing to explain a complex phenomenon by appealing to a metaphysical primitive invented just for the sake of this explanation sounds like a non-starter, especially if the primitive is ontological (that is, having to do with the basics of existence), as in the case of panpsychism. In any case, setting aside the question of the intuitive appeal of panpsychism, it seems not to have the explanatory value that is usually attributed to it (Goff, 2009). The alternative to panpsychism is emergence (Crutchfield, 1994; Van Cleve, 1990), which I discuss at some length in chapter 2, and which serves as the theoretical foundation for the present book. For further details, see the Dynamical Emergence Theory of phenomenal awareness (Moyal et al., 2020).

6. Recovery from general anesthesia can lead to weeks-long cognitive dysfunction, especially in the elderly (Bekker and Weeks, 2003).

7. The hole-in-the-skull argument is from (Merker, 2007).

8. The phrase "Minds are simply what brains do" is due to Marvin Minsky (1991).

9. The Identity Theory of the relationship between the mind and its instantiation in the brain has originally been developed in the 1950s (Feigl, 1958; Place, 1956); see also (Smart, 2004b). The version that I adopt in this book elaborates on the basic premise — that the mind *is identical to* what the brain does — by invoking the concept of emergence (Crutchfield, 1994; to be discussed in chapter 2).

10. Two highly entertaining and erudite treatments of the concept of computation are (Hofstadter, 1979) and (Rucker, 2016); see also my *Computing the Mind: How the Mind Really Works* (Edelman, 2008a).

11. To compute the sum of two numbers, the abacus is first cleared by pushing all the beads to the left. On the bottom-most string, two groups of beads, corresponding to the two numbers, are then slid to the right and the answer is read off (if there are not enough beads on the first string, the excess count is carried over to the next one).

12. See (Edelman, 2008b) or, for a book-length discussion, (Edelman, 2008a).

13. I have learned to appreciate the brain's sensory predicament by reading Merker (2007), who offers an unparalleled synthesis of vertebrate neuroanatomical and neurophysiological findings related to embodied consciousness; see also (Merker, 2012).

14. I wrote "its world," rather than "the world," because the animal's lived and experienced environment — what von Uexküll (2010) called its *Umwelt* — depends on the species and, indeed, is actively constructed by the members of the species under species-specific sensory, motor, and cognitive constraints.

15. The *E. coli* example is from (Tagkopoulos et al., 2008); see also (Lyon, 2015; Strassmann and Queller, 2011).

16. Associative learning in *E. coli* has been described by Tagkopoulos et al. (2008).

17. Friston et al. (2020) apply the Markov blanket formalism to the problem of other minds, which they introduce as follows: "Applying the concept of subjective, phenomenal consciousness to a system trivially presupposes that this system can be described from two perspectives (i.e., from a third- and from a first-person perspective)." Intuitively, the first-person perspective on the state of the system defined by a Markov blanket comprises the states that are directly accessible from within it; conversely, the third-person perspective is what can be known about these internal states from the outside, via the active variables (which, together with the sensory variables, form the Markov blanket).

18. Quoting from Friston's (2013, p.5) original invocation of the concept of Markov blanket in the context of cell biology: "[...] if we consider short-range electrochemical and nuclear forces, then a cell membrane forms a Markov blanket for internal intracellular states." The importance of the theoretical need to demarcate the inside of a cell from the outside is stressed by Godfrey-Smith (2019, p.10): "In a biological framework, one of the first things that can be explained is the existence of entities that are marked off from their surrounds in relevant ways."

19. Chalmers (1994) was the first to point out that consistent stepwise correspondence between sequences of internal and external states is astronomically less likely to happen by chance, compared to correspondence between a single external event and its purported internal representation. Smart (2004a, p.50) concurs: "Without a continuous miracle this would require a continuous succession of appropriate pure chance events and so its probability would be zero."

20. The evolutionary roots of representation are discussed in (Edelman, 2015). Dennett (1975) offers a popular exposition, involving a highly intuitive concept borrowed from computer science, the "Tower of Generate and Test."

21. The term "zombie" was introduced in a philosophical context in 1974 by R. Kirk and was subsequently popularized by Dennett (1991). In a recent review article, Kirk (2019) writes: "Few people, if any, think zombies actually exist. But many hold that they are at least conceivable, and some that they are possible." A 2013 poll of professional philosophers (D. Bourget and D. Chalmers, *What Do Philosophers Believe?*, unpublished article) yielded the following breakdown of opinions concerning zombies: conceivable but not metaphysically possible, 35.6%; metaphysically possible 23.3%; inconceivable 16.0%; other 25.1%. Personally, I cannot fathom why someone's ability to conceive of some state of affairs should make any difference whatsoever in any debate on any matter of fact. Unfortunately, such protestations carry very little weight, seeing that a person's *in*ability to conceive of something is equally inconsequential.

22. The simile comes from William James (1890, p.236): "When we take a general view of the wonderful stream of our consciousness, what strikes us first is the different pace of its parts. Like a bird's life, it seems to be an alternation of flights and perchings." Technically, this kind of dynamics is referred to as metastable: a regime that is in between stability, when the state does not change, and continuous change (Freeman and Holmes, 2005; Friston, 1997; Tognoli and Kelso, 2014). Edelman and Moyal (2017) discuss the computational reasons for metastability in the dynamics of behavioral control in general and consciousness in particular (see also Moyal and Edelman, 2019) .

23. The absolute necessity for consciousness to be intrinsic to the experiencer (rather than imputed from the outside; see p. 13) is entirely neglected by all but a few attempts at explaining it; the exceptions are the Integrated Information Theory (e.g., Oizumi et al., 2014; Tononi, 2008) and the Dynamical Emergence Theory (Fekete and Edelman, 2011; Moyal et al., 2020).

24. The utility of modeling the dynamics of the environment has been noted by Tagkopoulos et al. (2008, p.1313): "The widespread use of internal circadian models, from unicellular cyanobacteria to humans, suggests that this predictive mode of behavior confers considerable fitness advantages to organisms that have evolved it."

25. The importance of prediction — foresight and forethought — to all of cognition, and the resulting need to model the dynamics of the environment, is the central motif of (Edelman, 2008a). With regard to the *E. coli* example, Tagkopoulos et al. (2008, p.1316) write: "the correlation structure of the environment can be internalized as a probabilistic model in the high-dimensional space of an organism's complete sensory perception. As such, the very organization of microbial regulatory networks may, in large part, represent the physical instantiation of this probabilistic model."

26. See (Dennett, 1975); Dennett (1995) compares basic Darwinian "selection by death" with more sophisticated modes of learning, in which scenarios can

be evaluated before being tried out. In a later book (Dennett, 2003, p.53), he poses the key question: "How could a process with no foresight invent a process with foresight?" His detailed answer served as the basis for my present discussion.

27. According to Zahavi's (1975) Handicap Principle, a visible display of high fitness on part of an animal, intended to attract a mate or deter a predator, is only effective if it is made credible by an expenditure of energy that cannot easily be faked (the classical examples being the peacock's tail and antelope stotting). The Handicap Principle was later extended to cover honest signaling in general (Zahavi, 2018), leading to a theory that is widely seen as the greatest advance in evolutionary theory since Darwin.

28. This hypothesis regarding the evolutionary origins and computational nature of pain is due to Kolodny et al. (2021).

29. Reinforcement Learning or RL is an extremely general computational formalism: any learning problem can be formulated as an instance of RL (Schmidhuber, 2015). See (Sutton and Barto, 1998) for a comprehensive early introduction, (Littman, 2015) for the brain angle, and (Mnih et al., 2015) for an example of an extremely effective use of RL in artificial intelligence.

30. This property of computation is referred to as multiple realizability: the same recipe or algorithm can be implemented in vastly different substrates (Edelman, 2008a, p.162).

31. Cognitive information processing and feelings — notably, caring — are quite inseparable, as indicated by mounting evidence from psychology, neurobiology, and evolutionary science (e.g., Damasio and Carvalho, 2013; Feldman Barrett, 2013; Lindquist et al., 2012; Panksepp, 2001, 2005). For a glimpse of some of the relevant lines of reasoning and findings, consider how *valence*, which is an aspect of feeling, reduces to aversion and attraction, which are predispositions to action. Support for this notion comes from findings suggesting that approach-avoidance response is automatic. For behaviorists, the difficulty of disentangling liking from wanting and the likely futility of trying to do so (Havermans, 2011) are suggestive enough. In any case, it appears that "valence activates approach-avoidance behavior independently of evaluation and approach-avoidance intentions," as the title of (Krieglmeyer et al., 2010) proclaims; moreover, there is evidence of "an automatic pathway from evaluation to distance-change behavior" (Krieglmeyer et al., 2013). A study that combined transcranial magnetic stimulation with magnetoencephalography (TMS/MEP) found evidence for a direct link between positive/negative stimuli and approach/avoidance tendencies (Fini et al., 2020). Support for automatic and obligatory affective valuation of perceptual input was found by Turner et al. (2017, p.46), who report that "relevance of everyday objects is rapidly and automatically processed for a wide range of visual stimuli." More broadly, a meta-analysis by Beatty et al. (2016, p.237) suggests that "the central tenets of the biological predisposition, distance regulation, and cognitive evaluation accounts were supported." Another meta-analysis reached a comparable conclusion that the link between affect and approach-avoidance may be indirect and mediated by "appraisals or re-appraisals" (Phaf et al., 2014,

p.14). Finally, Eder et al. (2016, p.1) point towards "a theoretical integration of cognitive and motivational aspects as a more promising approach for future research."

References

Albahari, M. (2009). Witness-consciousness: Its definition, appearance and reality. *Journal of Consciousness Studies, 16*, 62–84.

Barlow, H. B. (1990). Conditions for versatile learning, Helmholtz's unconscious inference, and the task of perception. *Vision Research, 30*, 1561–1571.

Beatty, G. F., Cranley, N. M., Carnaby, G., & Janelle, C. M. (2016). Emotions predictably modify response times in the initiation of human motor actions: A meta-analytic review. *Emotion, 16*, 237–251.

Bekker, A. Y., & Weeks, E. J. (2003). Cognitive function after anaesthesia in the elderly. *Best Practice & Research Clinical Anaesthesiology, 17*(2), 259–272.

Chalmers, D. J. (1994). On implementing a computation. *Minds and Machines, 4*, 391–402.

Crutchfield, J. P. (1994). The calculi of emergence: Computation, dynamics, and induction. *Physica D, 75*, 11–54.

Damasio, A., & Carvalho, G. B. (2013). The nature of feelings: Evolutionary and neurobiological origins. *Nature Reviews Neuroscience, 14*, 143–152.

Dennett, D. C. (1975). Why the Law of Effect will not go away. *Journal of the Theory of Social Behaviour, 5*, 179–187.

Dennett, D. C. (1991). *Consciousness explained*. Little, Brown & Company.

Dennett, D. C. (1995). *Darwin's dangerous idea: Evolution and the meanings of life*. Simon & Schuster.

Dennett, D. C. (2003). *Freedom evolves*. Viking.

Edelman, S. (2008a). *Computing the mind: How the mind really works*. Oxford University Press.

Edelman, S. (2008b). On the nature of minds, or: Truth and consequences. *Journal of Experimental and Theoretical AI, 20*, 181–196.

Edelman, S. (2015). Varieties of perceptual truth and their possible evolutionary roots. *Psychonomic Bulletin and Review, 22*, 1519–1522.

Edelman, S., & Moyal, R. (2017). Fundamental computational constraints on the time course of perception and action. In C. Howard (Ed.), *Temporal sampling and representation updating*. Progress in brain research (Vol. 236, pp. 121–142). Academic Press.

Eder, A. B., Rothermund, K., & Hommel, B. (2016). Commentary: Contrasting motivational orientation and evaluative coding accounts: On the need to differentiate the effectors of approach/avoidance responses. *Frontiers in Psychology, 7*, 163. A commentary on Kozlik, J., Neumann, R., and Lozo, L. (2015). Frontiers in Psychology 6:563.

Evans, E. B., & Balster, R. L. (1991). CNS depressant effects of volatile organic solvents. *Neuroscience & Biobehavioral Reviews, 15*, 233–241.

Feigl, H. (1958). The 'Mental' and the 'Physical'. In H. Feigl, M. Scriven, & G. Maxwell (Eds.), *Concepts, theories, and the mind-body problem*. University of Minnesota Press.

Fekete, T., & Edelman, S. (2011). Towards a computational theory of experience. *Consciousness and Cognition, 20*, 807–827.

Feldman Barrett, L. (2013). Psychological construction: The Darwinian approach to the science of emotion. *Emotion Review, 5*, 379–389.

Fini, C., Fischer, M., Bardi, L., Brass, M., & Moors, A. (2020). Support from a TMS/MEP study for a direct link between positive/negative stimuli and approach/avoidance tendencies. *Neuropsychologia, 143*, 107496.

Freeman, W. J., & Holmes, M. D. (2005). Metastability, instability, and state transition in neocortex. *Neural Networks, 18*, 497–504.

Friston, K. (2013). Life as we know it. *Journal of The Royal Society Interface, 10*(86), 1–12.

Friston, K. J. (1997). Transients, metastability, and neuronal dynamics. *Neuroimage, 5*, 164–171.

Friston, K. (2018). Am I self-conscious? (Or does self-organization entail self-consciousness?). *Frontiers in Psychology, 9*, 579.

Friston, K. J., Wiese, W., & Hobson, J. A. (2020). Sentience and the origins of consciousness: From Cartesian duality to Markovian monism. *Entropy, 22*, 516.

Gasparri, L. (2019). Priority cosmopsychism and the Advaita Vedānta. *Philosophy East and West, 69*(1), 130–142.

Godfrey-Smith, P. (2019). Evolving across the explanatory gap. *Philosophical Theory and Practical Biology, 11*, 1.

Goff, P. (2009). Why panpsychism doesn't help us explain consciousness. *Dialectica, 63*(3), 289–312.

Goff, P., Seager, W., & Allen-Hermanson, S. (2017). Panpsychism. In E. N. Zalta (Ed.), *The Stanford encyclopedia of philosophy* (Winter 2017 ed.). Metaphysics Research Lab, Stanford University.

Havermans, R. C. (2011). "You say it's liking, I say it's wanting...". On the difficulty of disentangling food reward in man. *Appetite, 57*, 286–294.

Hofstadter, D. R. (1979). *Gödel, escher, bach: An eternal golden braid*. Basic Books.

James, W. (1890). *The principles of psychology*. Holt. Available online at http://psychclassics.yorku.ca/James/Principles/.

Kihlstrom, J. F. (1995). The rediscovery of the unconscious. In H. Morowitz, & J. Singer (Eds.), *The mind, the brain, and complex adaptive systems*. Santa Fe institute studies in the sciences of complexity (Vol. 22, pp. 123–143). Addison-Wesley.

Kihlstrom, J. F. (2022). Consciousness, the unconscious, and the self. *Psychology of Consciousness: Theory, Research, and Practice, 9*(1), 78–92.

Kirk, R. (2019). Zombies. In E. N. Zalta (Ed.), *The Stanford encyclopedia of philosophy* (Spring 2019 ed.). Metaphysics Research Lab, Stanford University.

Kolodny, O., Moyal, R., & Edelman, S. (2021). A possible evolutionary function of phenomenal conscious experience of pain. *Neuroscience of Consciousness, 7*(2), niab012.

Krieglmeyer, R., Deutsch, R., De Houwer, J., & De Raedt, R. (2010). Being moved: Valence activates approach-avoidance behavior independently of evaluation and approach-avoidance intentions. *Psychological Science, 21*, 607–613.

Krieglmeyer, R., De Houwer, J., & Deutsch, R. (2013). On the nature of automatically triggered approach-avoidance behavior. *Emotion Review, 5*, 280–284.

Lamme, V. (2015). The crack of dawn: Perceptual functions and neural mechanisms that mark the transition from unconscious processing to conscious vision. In T. Metzinger, & J. M. Windt (Eds.), *Open MIND* (Vol. 22). MIND Group.

Le Guin, U. K. (1969). *The Left Hand of Darkness*. Ace Books.

Lem, S. (1974). *The Cyberiad*. The Seabury Press. Translated by Michael Kandel.

Libet, B. (1985). Unconscious cerebral initiative and the role of conscious will in voluntary action. *Behavioral and Brain Sciences, 8*, 529–566.

Lindquist, K. A., Wager, T. D., Kober, H., Bliss-Moreau, E., & Feldman Barrett, L. (2012). The brain basis of emotion: A meta-analytic review. *Behavioral and Brain Sciences, 35*, 121–202.

Littman, M. L. (2015). Reinforcement learning improves behaviour from evaluative feedback. *Nature, 521*, 445–451.

Lyon, P. (2015). The cognitive cell: Bacterial behavior reconsidered. *Frontiers in Microbiology, 6*, 264.

Maoz, U., Yaffe, G., Koch, C., & Mudrik, L. (2019). Neural precursors of decisions that matter: An ERP study of deliberate and arbitrary choice. *eLife, 8*, e39787.

McCulloch, W. S. (1965). The past of a delusion. In *Embodiments of mind* (pp. 276–306). MIT Press. Presented to the Chicago Literary Club (January 28, 1952).

Merker, B. (2007). Consciousness without a cerebral cortex: a challenge for neuroscience and medicine. *Behavioral and Brain Sciences, 30*, 63–81.

Merker, B. (2012). From probabilities to percepts: A subcortical 'global best estimate buffer' as locus of phenomenal experience. In S. Edelman, T. Fekete, & N. Zach (Eds.), *Being in time: Dynamical models of phenomenal experience* (pp. 37–80). John Benjamins.

Metzinger, T. (2003). *Being no one: The self-model theory of subjectivity*. MIT Press.

Metzinger, T. (2004). The subjectivity of subjective experience: A representationalist analysis of the first-person perspective. *Networks, 3–4*, 33–64.

Metzinger, T. (2018). Minimal phenomenal experience. In *Third Carnap lecture, given on 9 March, 2018*. Ruhr-Universität Bochum.

Metzinger, T. (2020). Minimal phenomenal experience: Meditation, tonic alertness, and the phenomenology of "pure" consciousness. *Philosophy and the Mind Sciences, 1*(I), 7–44.

Metzinger, T. (2021). Artificial suffering: An argument for a global moratorium on synthetic phenomenology. *Journal of Artificial Intelligence and Consciousness, 8*(1), 1–24.

Minsky, M. (1991). Conscious machines. In *Machinery of consciousness*. Proceedings, National Research Council of Canada, 75th Anniversary Symposium on Science in Society.

Mnih, V., Kavukcuoglu, K., Silver, D., Rusu, A. A., Veness, J., Bellemare, M. G., Graves, A., Riedmiller, M., Fidjeland, A. K., Ostrovski, G., Petersen, S., Beattie, C., Sadik, A., Antonoglou, I., King, H., Kumaran, D., Wierstra, D., Legg, S., & Hassabis, D. (2015). Human-level control through deep reinforcement learning. *Nature, 518*, 529–533.

Moyal, R., & Edelman, S. (2019). Dynamic computation in visual thalamocortical networks. *Entropy, 21*(5), 500.

Moyal, R., Fekete, T., & Edelman, S. (2020). Dynamical emergence theory (DET): A computational account of phenomenal consciousness. *Minds and Machines, 30*, 1–21.

Oizumi, M., Albantakis, L., & Tononi, G. (2014). From the phenomenology to the mechanisms of consciousness: Integrated Information Theory 3.0. *PLoS Computational Biology, 10*(5), e1003588.

Panksepp, J. (2001). The neuro-evolutionary cusp between emotions and cognitions: implications for understanding consciousness and the emergence of a unified mind science. *Evolution and Cognition, 7*, 141–163.

Panksepp, J. (2005). Affective consciousness: Core emotional feelings in animals and humans. *Consciousness and Cognition, 14*, 30–80.

Phaf, R. H., Mohr, S. E., Rotteveel, M., & Wicherts, J. M. (2014). Approach, avoidance, and affect: A meta-analysis of approach-avoidance tendencies in manual reaction time tasks. *Frontiers in Psychology, 5*, 378.

Place, U. T. (1956). Is consciousness a brain process? *British Journal of Psychology, 47*, 44–50.

Rinaldi, A. (2014). Reawakening anaesthesia research. *EMBO Reports, 15*(11), 1113–1118.

Rosenthal, D. (1990). A theory of consciousness. Research Group on Mind and Brain, ZiF Report No. 40, University of Bielefeld.

Rosenthal, D. M. (2009). Higher-order theories of consciousness. In A. Beckermann, B. P. McLaughlin, & S. Walter (Eds.), *The Oxford handbook of philosophy of mind* (Chap. 13, pp. 239–252). Oxford University Press.

Rucker, R. (2016). *The lifebox, the seashell, and the soul* (2nd ed.). Transreal Books.

Rudolph, U., & Antkowiak, B. (2004). Molecular and neuronal substrates for general anaesthetics. *Nature Reviews Neuroscience, 5*, 709–720.

Schmidhuber, J. (2015). Deep learning in neural networks: An overview. *Neural Networks, 61*, 85–117.

Smart, J. J. C. (2004a). Consciousness and awareness. *Journal of Consciousness Studies, 11*, 41–50.

Smart, J. J. C. (2004b). The identity theory of mind. In E. N. Zalta (Ed.), *Stanford encyclopedia of philosophy*. Stanford University. Available online at http://plato.stanford.edu/archives/fall2004/entries/mind-identity/.

Sperry, R. W. (1969). A modified concept of consciousness. *Psychological Review, 76*, 532–536.

Sperry, R. W. (1970). An objective approach to subjective experience: Further explanation of a hypothesis. *Psychological Review, 77*, 585–590.

Strassmann, J. E., & Queller, D. C. (2011). Evolution of cooperation and control of cheating in a social microbe. *Proceedings of the National Academy of Science, 108*, 10855–10862.

Sutton, R. S., & Barto, A. G. (1998). *Reinforcement learning*. MIT Press.

Tagkopoulos, I., Liu, Y.-C., & Tavazoie, S. (2008). Predictive behavior within microbial genetic networks. *Science, 320*, 1313–1317.

Tognoli, E., & Kelso, J. S. (2014). The metastable brain. *Neuron, 81*, 35–48.

Tononi, G. (2008). Consciousness as integrated information: a provisional manifesto. *The Biological Bulletin, 215*, 216–242.

Turner, W. F., Johnston, P., de Boer, K., Morawetz, C., & Bode, S. (2017). Multivariate pattern analysis of event-related potentials predicts the subjective relevance of everyday objects. *Consciousness and Cognition, 55*, 46–58.

Van Cleve, J. (1990). Mind: Dust or magic? Panpsychism versus emergence. *Philosophical Perspectives, Action Theory and Philosophy of Mind, 4*, 215–226.

von Helmholtz, H. (1856). Unconscious conclusions. In W. N. Dember (1964) (Ed.), *Visual perception: The nineteenth century* (pp. 163–170). Wiley.

von Uexküll, J. (1934/2010). *A Foray into the world of animals and humans: With a theory of meaning*. University of Minnesota Press. Translated by J. D. O'Neill.

Waldron, W. S. (2002). The dependent arising of a cognitive unconscious in Buddhism and science. *Contemporary Buddhism, 3*, 141–160.

Zahavi, A. (1975). Mate selection: A selection for a handicap. *Journal of Theoretical Biology, 53*, 205–214.

Zahavi, A. (2018). Animal communication: The handicap principle. In H. Callan (Ed.), *The international encyclopedia of anthropology*. John Wiley & Sons.

2 Minimal Selves

Tired starlings. The web of cause and effect. Emergence: the effective self as a local causal domain. Emergence and causality: Where has all the physics gone? The self and others. The captive scapegoat and the birth of tragedy. The second revolution.

Rien ne réussit comme le succès.
(Nothing succeeds like success.)

Ange Pitou
— ALEXANDRE DUMAS

Tired Starlings

MINIMAL phenomenal experience, from which the tower of self-consciousness grows out when the occasion calls, is a peculiar foundation to build on. Like everything else about consciousness, it is dynamic: not a state but a process. As such, it is best referred to not by a noun but by a verb: ideally, something that suggests swarming or flocking.[1]

The murmuration of feelings that together comprise minimal conscious experience arises from the interactions of the cognitive processes that I called the Gang of Four: the memory-based predictor, the actor posse, the decision maker, and the critic. Evolution ensures that these computations, which are experienced emotionally, have their various uses: they are "algorithms for survival."[2]

Computation that is intrinsically emotional and care-laden by its very nature — that is, by virtue of the self-interpreting dynamics of the representation-space trajectory — takes a toll on the system that implements it. Feelings are tiring, often more so than physical exertion; and psychache can be as unbearable as physical pain, and more deadly. It could not have been otherwise: were emotions not unavoidably felt, sometimes strongly so, they would not be effective as tools of behavioral self-control. And so it goes: evolution's propensity to breed creatures capable of effective behavioral control brought about the second consciousness revolution — the inception of the self.

S. Edelman, *The Consciousness Revolutions*,
https://doi.org/10.1007/978-3-031-24012-6_2

The Web of Cause and Effect

The seed of the conscious self, along with all the implications its existence has for the human condition, is contained in the very concept of effectiveness. The essential property of effectiveness is that it is relative: it hinges on whether or not doing *this* beats doing *that* (by how much or in what respects is less important). Having this kind of assessment scale in place is what makes *learning* possible: instead of endlessly flailing about while trying one random move after another, you can flail once, assess damage, find out what went wrong, aim to do better, and repeat. This method, which will always work as long as the damage is not fatal, is precisely the generic recipe for reinforcement learning, which I first brought up in chapter 1.

In all learning that uses past outcomes to improve future behavior, there are two main tasks: to determine what went wrong and to figure out how to set it to rights. Both these tasks can be very challenging, due to the key property of causality: in this universe, every event that is of potential interest to any sufficiently complex autonomous system, such as a microbe or a marmot, has both multiple causes and multiple consequences. Together, such "local" events form a web of cause and effect, which is part of a greater web — one that encompasses the entire universe.[3]

Because the structure of the great web of cause and effect is ultimately responsible for the advent of selves, it merits a closer look. The first thing to notice is that the great web is all-encompassing in the sense that an event that is left out of it — that is, one that has no causes and no consequences — is effectively nonexistent, as far as this universe is concerned.[4] Next, consider events that have no causes. In physics, such events are allowed by some theories, but not by others. Quantum mechanics, in particular, holds certain radioactive decay events to be truly and irreducibly random (thus impossible to anticipate except statistically) and therefore lacking a cause.[5]

A random and uncaused event can, however, have consequences, so it does have its place within the great web. What about an event that has no consequences? Such a causal "dead end" leaves no trace after it. Because the fact that it has transpired makes no difference to the rest of the universe, it might as well not exist. Lastly, an event that has only a single consequence can be effectively merged with it. Likewise, an event that has only a single cause can be merged with its antecedent.[6] Therefore, such singly connected events need not be considered in what follows.

Felix qui potuit rerum cognoscere causas (Happy the Man, who, studying Nature's Laws, / Thro' known Effects can trace the secret Cause) —Virgil (and Dryden's translation)

This quick analysis supports the notion that in the great web of cause and effect any event of interest is the nexus of multiple converging and diverging causal chains. This state of affairs has repercussions for decision-making and for learning. The chief one of these is the so-called Credit Assignment Problem — a well-known and thoroughly researched topic in the computational theory of learning.[7] Briefly, since for an event that is embedded in the web of cause and effect there exists no unique antecedent, there is no principled way for a learner to single out *the* past decision — either to act or, counterfactually, to refrain from action — to credit or blame for its outcome. This leads to an exponential explosion in the number of scenarios that the learner must consider. Exponential (and therefore prohibitive) computational complexity is encountered also if the antecedent's credit, whether positive or negative, is permitted to be graded rather than all-or-none. Moreover, the same applies not only to the antecedents of an event but also to its consequences, which ramify exponentially.

Despite this potential for being unmanageably complex, learning is in fact tractable — perhaps not always, but often enough for life to come into being and even to evolve into creatures that ponder the complexity of learning. For that, we can thank the physics of causation. There is one aspect of the structure of the great web of cause and effect that can be put to use to make learning happen: the natural sorting of events into "causal neighborhoods," such that only those events that are in some sense near the locus of learning need to be taken into account.

One way to delineate a causal neighborhood is by appealing to the concept of light cone, which is central to Einstein's special relativity theory (SRT). Because SRT rules out faster-than-light propagation of information, all events causally related to a given one must reside within a sector of spacetime that has the shape of a double cone, which extends backward and forward in time and whose apex is at the event in question.[8] Having to consider only those potential antecedents that fall within the given event's past light cone does reduce somewhat the required legwork.

Alas, the light cone structure by itself is insufficiently constraining to ensure the tractability of tracking cause and effect. Light is so fast that opening even a small time window yields a huge chunk of space (about $300,000$ kilometers of it for each second), which may contain too many potential events to consider. A much stronger constraint becomes apparent when events within the light cone are further sorted by the strength of their possible interaction with the event at the apex, as measured by the energy that is exchanged. A low-energy interaction can be illustrated by a micrometeorite that burns up in the stratosphere and is seen by you, causing you to make a wish — an outcome whose further consequences are likely to peter out, resulting in a causal dead end. In comparison, a high-energy interaction, with a correspondingly larger effect on your subsequent life, might be getting hit between the eyes by a pebble of the same size, even if it comes from a slingshot and not the interplanetary space. If looks could kill, David would merely have to show Goliath his sling, instead of using it.

Emergence: The Effective Self as a Local Causal Domain

An interaction that is energetically weak implies that the interacting regions of spacetime belong to effectively distinct *causal domains*. A dynamical system that is initially homogeneous or unstructured, in that it possesses no intrinsic subdivisions, may then with time differentiate into subsystems, such that the coupling between those is significantly weaker than the coupling within each of them (in the technical sense of the amount and rate of energy exchange).

Such differentiation of a whole into parts, which takes place gradually as the system's dynamics plays out, exemplifies a key concept in the science of complex systems: *emergence*, or the appearance of structure where previously there has been none:

> Emergence is generally understood to be a process that leads to the appearance of structure not directly described by the defining constraints and instantaneous forces that control a system. Over time "something new" appears at scales not directly specified by the equations of motion. An emergent feature also cannot be explicitly represented in the initial and boundary conditions. In short, a feature emerges when the underlying system puts some effort into its creation.

James Crutchfield, from whose pioneering work on emergence I have borrowed this exposition, notes that the emergent features can only be "new" in the eye of an observer. Crucially, he proceeds to point out that this role can be played by the system itself, which makes it "intrinsic emergence: the system itself capitalizes on patterns that appear."[9]

As a central explanatory concept in complex systems theory, intrinsic emergence plays a key role in theoretical biology, in understanding life; and in mind science, in understanding consciousness.[10] However, the concept of emergent causal domains applies also at much lower levels of organization of matter. In particular, the presence of local causal domains characterizes certain states of condensed matter (as opposed to high-energy plasma).[11] In solids, in particular, "local" domains arise on multiple spatial scales. At small scales, quarks are bound by the strong nuclear force inside hadron particles; hadrons are bound by the residual strong force inside atomic nuclei; and atoms are bound by the electromagnetic force into molecules. At intermediate scales, local domains that are essential to the existence of life originate as assemblies of molecules that self-organize into stable superordinate structures, from single-cell organisms such as the *E. coli* bacterium to multicellular ones. Finally, at larger scales, there are interactions between organisms and their environment, both animate and inanimate.

A living cell is a paradigmatic example of an emergent local causal domain. In a prokaryote cell, a physical barrier, the cell membrane, shuts out the universe, making metabolic processes local and therefore controllable (in eukaryotes, there is an additional barrier that defines the nucleus; also organelles). Because the membrane is selectively partly permeable, processes that are internal to the cell can be affected also by outside events. Given that the cell's sensors are all chemical or mechanical, outside events too must be local in order to have a non-negligible effect on the internal processes and therefore on the cell's conduct.

Two subsequent evolutionary developments in multicellular organisms greatly extended the reach of perception: photosensitive cells, many millions of which are found in the human retina; and a lens-equipped camera-like eye, capable of forming a crisp image of a distant object on the retina. In a photoreceptor, the cell membrane is adapted to sensing light: the inside-outside voltage difference that the cell maintains can be affected by the passage of a photon.[12] Because the photon may have traveled for light-years before causing the discharge, it may seem that using sensitive photoreceptors and good optics would destroy the ability to identify and learn from the causes of sensory events. How can you learn if photons signifying an event of notice may come from anywhere within your local galaxy?

Sophisticated optical sensing could have indeed been bad news for learning, but it isn't. Learning requires more than sensing: if a sensed event has no discernible consequences within a reasonable window of time, it is let go. If, on the other hand, it is implicated in an effect that is local in the mechanical/energetic sense discussed earlier, then it is interpreted as a possible cause and taken note of for future reference.[13] Immediate locality thus remains important: a system that learns only from energetically significant outcomes can easily come to distinguish between the inconsequential twinkling of Proxima Centauri in the southern sky and a potentially very consequential glimpse of a jaguar in the undergrowth. Likewise, a Goliath that survives his first encounter with David will the next time around at least consider standing down at the mere sight of a loaded sling.

Our notion of locality therefore stands. If parts of the great web of cause and effect that are physically individuated as local domains are sufficiently compact, tracing the likely antecedents and consequences of an event becomes computationally feasible. Under such conditions, evolution is poised to favor emergent structures that discover and use reinforcement learning, subject to the following two-part Locality Principle:[14]

The LOCALITY PRINCIPLE for reinforcement learning: local causes and local effects take precedence.

The Locality Principle:

(i) in assigning credit/blame for a local outcome, seek energetic local causes first (which may be portended by optical signs visible from far away);

(ii) in predicting outcomes for a local cause, examine local effects first.

Reinforcement learning that adheres to the Locality Principle, which is underwritten by the fundamental physics of the universe, is more likely to be tractable. This, in turn, promotes the evolution of autonomous systems that learn to deal with the outside world: nothing succeeds like success.[15]

INDIVIDUALS are aggregates that preserve a measure of temporal integrity, i.e., "propagate" information from their past into their futures.
— Krakauer et al. (2020)

Importantly, the concept of "outside" is now seen as potentially having not only a physical, but also a functional — that is, computational — meaning. In the literal sense, a system having an outside means that there exists a physical barrier that encloses its inside, in the same manner that a cell membrane encloses its protoplasm. Independently of that, a set of interacting objects, which may or may not be surrounded by a physical enclosure, can develop a "functional outside": all other objects, with which the founding members interact but weakly.

This is precisely what happens every time there emerges a distinct *individual*: a self-organizing, persistent local causal neighborhood, in relation to which the rest of the world is outside. The grouping of objects within such a neighborhood is *virtual* in the sense that it is due to ongoing computational activity, rather than to some fixed properties shared by its constituents. This virtual entity, whose existence nevertheless has real internal and external consequences, is what I call the *minimal effective self*.

Emergence and Causality: Where Has All the Physics Gone?

The minimal self plays a key role in conscious systems: it is the core component of *agency*, or the capacity for initiating actions and for taking responsibility for their outcomes. Concerning outcomes, the minimal self makes effective learning possible on part of for the system from which it emerges, by serving as the lightning rod for credit and blame assignment. As to action initiation, the process to which agency is attributed serves as that level in the system's organizational structure where the proverbial explanatory buck stops.

The perennial question "Why did the chicken cross the road?" can be addressed on many levels, of which one stands out in its cogency. It is the personal-agentic level, where the answer is "Because it wanted to." Crucially, this level is also the one that is indispensable for the system itself, in this case the chicken, which may wonder "Why did I cross the road?"[16] Dragging in brain science, such as it may be ("My brain made me do it"), biochemistry, or subatomic physics simply would not do. This is because blaming some neurons, or a metabolic pathway, or quarks and leptons would not help *me* to deal with being nearly run over by a speeding car while stepping down from the curb. *I* crossed the road because *I* wanted to; and without a minimal self, there can be no *I*.

To better appreciate the applicability in such cases of multiple possible levels of explanation, of which only one is truly to the point, we may consider a chess-playing computer. Reasons for its making a particular move can be found in its knowledge of the rules of chess, in the functioning of its electronic components, and in the physics that underlie that functioning. By analogy to the case of the chicken, though, it is only the rules of chess that offer a relevant explanation here. This is because the same chess game state (board configuration) or piece movement can be represented on the level of electronics by a large (in fact, infinite) number of physical states, such as voltages and currents. In other words, even a fully specified state transition on the level of electronics (let alone basic physics) does not explain why the computer decided to move this piece and not that one.

Paradoxically, it is disregarding — abstracting away — the lower-level details that makes effective explanation possible.[17] Generally speaking, in the hierarchy of levels of description that applies to every condensed-matter system,[18] the level on which agentic, learning selves are found is that of coarse-grain *macrostates*: an emergent subdivision of the system's state space, such that all the low-level or fine-grain states within each resulting region are functionally equivalent.[19]

The Self and the Others

The intrinsic — that is, internally emergent, rather than externally imposed — aggregation of causes and consequences of events into an effective self is bound to be widely replicated: where there is one self, there are likely to be many. Of course, not all the selves that emerge out of the great web of cause and effect interact with each other: those that do, do so because of their co-location. The great web thus possesses a hierarchical structure, somewhat resembling the hierarchy of levels of organization of condensed matter, which

I mentioned just now. The difference is that, unlike the physics of matter, the organization of the virtual effective selves is virtual too.

Mathematically speaking, an ensemble of interacting selves is a collection of dynamical systems with weak mutual coupling (as compared to the relatively stronger internal coupling of each system's constituent elements).[20] What sustains this structure is the ongoing computation carried out by the parties involved. First, the constituent elements of an effective self hang together only insofar as they act on each other. Second, they only act on each other because the ongoing process of reinforcement learning identifies them (for each other, not for any outside entity) as local and therefore selects them as the targets of interventions that are likely to improve outcomes. The key kind of successful outcome here is the continued virtual existence of the self.

The existence of other selves can greatly complicate life for the learner. The inner workings of other selves are mostly masked from outside scrutiny and intervention. On the one hand, the possibility of having "inner life," insulated from outside meddling, is an important advantage of being an effective self (just like in *E. coli* the membrane shields its metabolism and cognition from being drowned in noise). On the other hand, however, the masking makes it impossible for an agent to do a thorough job understanding the behavior of others and adjusting one's own behavior to anticipate or to make up for it.

In natural living ecosystems, the interactions of agents among themselves and with non-agentic factors (weather, climate, etc.) can be exponentially more complex than in a toy situation. An ecosystem is not, however, an entity that needs to plan its behavior. Its dynamics is what it is and if some individual or species doesn't like it, tough luck. In contrast, agents that are embedded in an ecosystem do have a problem: how not to get swept up in that mindless dynamics. The scale of this problem grows with the complexity of the individual agents' selves, and with their number.

This is where humans shine: while even bacteria can and do have non-trivial social life, human group dynamics really knocks it out of the park. This is why, once neurobiology and psychology are recognized as being far more complex than physics, sociology must be acknowledged as the ultimate challenge to science as we know it — merely because living as part of a society is the ultimate challenge for the individual.

The nature of human sociality differs greatly from that of other highly social species, such as bees, or termites. A bee is as sentient as we are,[21] and it cares about the well-being of the hive, but it has no sentiments that are specifically what we call *moral* — no ambition or envy, no guilt or shame, no gloating or schadenfreude. These emotions are all peculiarly human because they arise out of our peculiar sociality. Not only do our morals depend on language for full development and impact (as we shall see in chapter 4) — unlike the bees in a hive, we are genetically and culturally diverse. It is thus only to be expected that our cooperation should always be qualified and conditional, often on processes whose complexity boggles the mind, and our morals — often a morass.

The Captive Scapegoat and the Birth of Tragedy

It is unfortunate that the one aspect of the human self's existence that seems to be the most emotionally charged — how to live with others — is also the one that is computationally the most difficult to manage. The core of the problem is, however, not complexity as such, and it arises, independently of any social interaction, already within an individual self. The problem is this: caring about outcomes (rather than merely being more likely to choose some over others), along with the occasional negative affect that ensues from caring, is impossible to cancel at will.

SUFFERING is a state of negative affect that cannot be escaped just because it is disliked.

By definition, *suffering* is what we call a state of negative affect that cannot be escaped just because it is disliked.[22] It was the emergence of selves from the great web of cause and effect that paved the way for this new mode of existence. The whole point of booting up the effective self is to promote learning by making credit assignment tractable, and where there is credit, there is also blame.[23] Reinforcement learning cannot proceed without the critic process — but what use is a critic if failure is nobody's fault?

The self is that, which is to be blamed; that, which can be faulted for failures, big or small. It is a virtual computational entity created by the system especially to serve as a clearing house for credit and blame, which are channeled to it by the critic process. By also giving the self global access[24] to all the resources needed for deciding the course of action, the self-conscious system becomes more effective both at learning and at using what has been learned.

Suffering is the price that is paid for this effectiveness. Just like the feelings of valuation and caring, which play a key role in minimal consciousness, have evolved to be closely tied to action tendencies and therefore difficult to resist, the effective self has evolved to be a complex feeling, a particular kind of experience that cannot be easily set aside. In accordance with the two-part role of the self in receiving credit/blame and in steering behavior, that experience is twofold: it combines the feeling of being responsible with the feeling of being in control.[25]

As the designated receptacle for blame, the self is the system's scapegoat. Unlike the biblical scapegoat, which on the High Holy Days was loaded with the people's sins and driven off away from Jerusalem and into the desert, the self is held captive for the entire duration of its virtual existence. This existence is frequently interrupted: even when we are awake, much of our cognitive processing happens without involving the self ("subconsciously"). At the same time, every deviation from the routine, and every crisis, no matter how minor, causes the system to jerk the self's tether, wake it up, and bring it online.[26]

Admittedly, it's not all thorns all the time for the captive scapegoat: occasionally, it allows itself a virtual carrot of happiness, conjured by computational magic out of the same thin air. Because reward is ineffective — and thus, as far as evolution is concerned, meaningless — unless coupled with punishment, the happy state cannot persist for long. The carrot is just what it is per the popular cliche: a trick, a ploy to make the goat work harder. Our happiness, such as it is, is the happiness of pursuit.[27]

This, then, is the tragic sense of the life of the self: to be brought into existence just to be goaded by fleeting rewards to work harder, and to suffer when underperforming.[28] The suffering is always there: either present, as when self-punishment is under way, or anticipated, as when happiness is revealed, upon reflection, to be fleeting. That the minimal, "scapegoat" self, which lacks the capacity for higher-order conscious reflection and language, cannot describe its predicament in so many words hardly diminishes the tragedy.

The Second Revolution

The emergence of effective selves — coherent and persistent local causal domains in the great web of cause and effect — is the second consciousness revolution. It has been accompanied by the invention of suffering: a category of conscious states characterized by a combination of the feelings of being in control and of negative affect, which the newly emergent computational construct, the self, cannot cancel at will.

Both these characteristics are necessary for suffering to fulfill its evolutionary role. Effective behavior needs both an outlet for blame and a fulcrum for mounting action (which in turn can generate blame); the self serves as both.[29] Moreover, although the self is a virtual entity — like all feelings, that of being a self boils down to certain transient properties of the dynamics of the system out of which it arises — to itself it is more real than anything else in the universe. This is why there is nothing virtual about suffering: after all, serving as a vessel for suffering is to a large extent what the self is for.

Is there any hope of alleviating this existential suffering, or perhaps even abolishing it altogether? I shall return to this rather significant question in later chapters. We have, however, already taken the first steps towards developing a credible approach to it, by recognizing the evolutionary origins of suffering and shedding some light onto its computational nature.

❧ ❧ ❧

Notes

1. Concerning noun and verb roles in language, see *Tlön, Uqbar, Orbis Tertius* by Borges (1970):

> There are no nouns in Tlön's conjectural *Ursprache*, from which the "present" languages and the dialects are derived: there are impersonal verbs, modified by monosyllabic suffixes (or prefixes) with an adverbial value. For example: there is no word corresponding to the word "moon," but there is a verb which in English would be "to moon" or "to moonate." "The moon rose above the river" is *hlor u fang axaxaxas mlo*, or literally: "upward behind the onstreaming it mooned."

2. "Algorithms for survival: a comparative perspective on emotions" is the title of (Bach and Dayan, 2017). Borges (1999) includes a list of emotions, quoted from "testimonial ... from the third part of Torres Villarroel's *Vida e Historia*: "I am angry, fearful, compassionate, joyous, sad, greedy, generous, enraged, meek, and all the good and bad emotions and all the praiseworthy and reprehensible actions that can be found in all men together or separately." The multidimensional analysis of Cowen and Keltner (2017) yielded 27 emotions, joined by "continuous gradients."

3. The view of the universe as a great web of cause and effect is derived from the principle of dependent origination, which is central to Buddhism both as a philosophical system and as a religion. With regard to the former, see (Garfield, 1994) for an illuminating discussion of this principle and its relation to the doctrine of emptiness. The latter aspect is illustrated by this line from the Maha Hatthipadopama (Great Elephant Footprint) Sutra: "Whoever sees dependent co-arising sees the Dhamma [the Way]" (translated from Pāli by Thanissaro Bhikkhu). A link to Western philosophy has been offered by Scharfstein (1998, p.464):

> [David] Hume is clearly the most Buddhistic of the three English philosophers. This is in part because he is doubtful of the existence of God, but mainly because he denies — in rational theory though, quite un-Buddhistically, not in living emotion — the existence of the self. Using the equivalent of the Buddhist term *heaps* (skandhas), Hume ventures that human beings are only bundles or collections of different perceptions that succeed one another in rapid, perpetual flux. His general equivalent to the Buddhist law of dependent origination is defined by the set of rules by which we recognize and anticipate cause and effect.

4. Here is physicist Carlo Rovelli's (2021, p.76) remark on the interconnectedness of the causal universe: "If there was an object that had no interactions, no effect upon anything, emitted no light, attracted nothing and repelled nothing, was not touched and had no smell... it would be as good as nonexistent. To speak of objects that never interact is to speak of something — even if it existed — that could not concern us. It is not even clear what it would mean to say that such objects "exist.""

5. In contrast, superdeterminism posits all events to be deterministic, with causes that possibly go all the way back to the Big Bang. See ('t Hooft, 2016).

6. Cf. (Smolin, 2020, 3.1.2): "No elementary event has only one cause."

7. The Credit Assignment Problem was first identified by Marvin Minsky (1961, p.20).

8. In contrast, Lee Smolin's (2018; 2020) foundational theory of cosmology holds causal proximity to be logically prior to spacetime, which emerges from it.

9. The quoted definition of emergence is from (Crutchfield, 1994, p.12). See also (Kauffman and Clayton, 2006) on "self-propagating organization." A philosophy of neuroscience perspective on the "consciousness revolution" underwritten by emergence can be found in (Sperry, 1987).

10. As its name indicates, the Dynamical Emergence Theory of consciousness (Moyal et al., 2020) aims to show how minimal phenomenal consciousness can emerge from a computational process with the right kind of dynamics.

11. Condensed matter states are characterized by lower average energy compared to plasma. Smolin (2018, p.123) explicitly invokes energy in his calculus of events: "[...] the locality of distant events is dependent both on the observer and the energy carried by the probe that observer uses to image the event." On a related note, Gärdenfors (2020) bases his theory of events on geometry and force.

12. Because the probability of a photon being caught during a single passage through the membrane is small, the "business end" of a photoreceptor cell has the membrane folded many times over. In addition, in nocturnal animals such as cats, the photon catch efficiency is further boosted by the eye having a mirror-like innermost layer, which reflects a large proportion of the photons that make it that far in right back at the photoreceptors.

13. MacIver (2009) discusses the revolution in non-local sensing that was made possible by the emergence of visual systems based on imaging — which requires optics, such as a lens or at least a pinhole, and not just a photosensitive element.

14. Note that as far as the physics behind this Locality Principle is concerned, it does not matter whether locality is defined according to Einstein or according to Smolin (see note 8 above). As to the contemporary computational treatment of locality, Fano et al. (2016, p.137) point to the origins of this concept in the Locality Property introduced by Gandy (1980, p.135):

> In Turing's analysis the requirement that the action depend only on a bounded portion of the record was based on a human limitation. We replace this by a physical limitation which we call *the principle of local causation.* Its justification lies in the finite velocity of propagation of effects and signals: contemporary physics rejects the possibility of instantaneous action at a distance.
>
> > **Principle IV** (Preliminary version). *The next state, Fx, of a machine can be reassembled from its restrictions to overlapping "regions" s and these restrictions are locally caused. That is, for each region s of Fx there is a "causal neighborhood" $t \subseteq TC(x)$ of bounded size such that $Fx \upharpoonright s$ depends only on $x \upharpoonright t$.*

It should be noted that Gandy's machines are defined in terms of labeled hereditary finite sets. For the definition of the set restriction operation in that context, denoted by $\upharpoonright$, see Gandy (1980, p.128, (3)). Intuitively, $x \upharpoonright A$ is the structure whose construction follows that of x omitting all labels not in A.

15. In evolutionary science, the phrase "nothing succeeds like success" has been invoked *as an explanatory principle* by T. H. Morgan (1910, p.210) and Ernst Mayr (1976a, p.34); elsewhere, Mayr (1976b) explains why there is no tautology involved. This view of evolution is related to the anthropic principle (Wallace, 2019) and to the hypothesis of cosmological selection (Giudice et al., 2019; Smolin, 1997).

16. An accessible discussion of the levels of analysis methodology, which is central to mind science, can be found in (Edelman, 2008, section 4.2), with the chicken road-crossing example discussed on p.70. The classical source is (Marr and Poggio, 1977).

17. The paradox of a higher-level explanation superseding a lower-level one is typical of the so-called "special sciences." A chemical reaction is best explained by appealing to chemistry, not subatomic physics; likewise, the functioning of a living cell is best explained in the language of biology, not chemistry or physics. Ross and Spurrett (2004) argue convincingly that all sciences except fundamental physics are special in this sense (which is why it is physics that should be called special). A solution to the problem of multiple-level causation, called contextual emergence, is offered in (Atmanspacher, 2017; Harbecke and Atmanspacher, 2012). The chess example is from the short paper on agency by Rovelli (2020), who concludes it by remarking that "Ultimately, agency is *always* nothing else than ignoring some physical links."

18. The ubiquity of hierarchical organization in physics etc. has been famously remarked upon by Simon (1973, p.26):

> Scientific knowledge is organized in levels, not because reduction in principle is impossible, but because nature is organized in levels, and the pattern at each level is most clearly discerned by abstracting from the detail of the levels far below. [...] And nature is organized in levels because hierarchic structures — systems of Chinese boxes — provide the most viable form for any system of even moderate complexity.

19. Crutchfield et al. (2014, p.2) defines a macrostate as follows: "A macrostate — say, given by particular values of temperature and pressure — is the set of the many microscopic molecular configurations or microstates that lead to the same measurable properties." Equivalent definitions can be found elsewhere (Atmanspacher, 2016; Shalizi and Moore, 2003). Allefeld et al. (2009) show how macrostates can be identified in recordings of brain activity.

20. Yoshimi (2012) discusses the relevance of notion of coupling in dynamical systems to cognitive science.

21. Edelman et al. (2016) outline an argument for bee consciousness.

22. Calling suffering "the cognitive scotoma" because of how rarely it is discussed by philosophers, Metzinger (2017, p.244) writes: "We lack a comprehensive theory of conscious suffering. One of the key desiderata is a conceptually convincing and empirically plausible model of this very specific class of phenomenal states: those that we do not want to experience if we have any choice, those states of consciousness which folk-psychology describes as "suffering"."

23. The asymmetry between credit and blame is vividly illustrated by this comment, made by Galeazzo Ciano, Mussolini's son-in-law, who served as

foreign minister of Italy until 1943, when he was executed on his father-in-law's orders: "Come sempre, la vittoria trova cento padri, e nessuno vuole riconoscere l'insuccesso" ("As always, victory finds one hundred fathers, and none wants to recognize failure"; diary entry for September 9, 1942).

24. Global access is the defining characteristic of consciousness according to the Global Workspace Theory (GWT) proposed by Baars (e.g., 2005) and developed by others (e.g., Dehaene et al., 2014; Shanahan, 2010). While related ideas have been put forward by others (e.g., the "global model of neuronal command-control" of Llinás and Iberall, 1977, and the "global best estimate" of Merker, 2012), the GWT stands out in its claim that global access to information suffices to account for phenomenal experience. This has prompted Frith and Metzinger (2016, p.201) to ask the obvious question:

> For many activities there is clear need for "global availability" of information. But why should this global access be associated with subjective experience? Furthermore, decisions, which require the integration of many sources of information, seem to be better made in the absence of conscious reflection [...] perhaps because appropriate weighting of multiple sources of information is disrupted by conscious deliberation[.]

Indeed, a recent imaging study (Mei et al., 2022) found the purported signature of "global access" to be associated with unconscious rather than conscious processing of visual stimuli. I should add that, according to the approach developed in this book, discernment-supporting representations may remain unconscious simply because they are ignored by, or not available to, the brain process that has current access to the reporting outlet such as language or overt voluntary action (such as button-pressing).

25. We must pay close attention to language here, to avoid the conceptual confusion that is inherent in the common usage of key terms. It makes no sense to talk about "having a self": who is it that has it? One does not *have* a self — one *is* a self.

26. Thus, (Smith et al., 2003, p.338): "If you watch an aging cat consider a doubtful leap onto the dryer, you will suspect that what James (1890, p.93) said is true, 'Where indecision is great, as before a dangerous leap, consciousness is agonizingly intense.' "

27. "The Happiness of Pursuit" is the title of (Edelman, 2012).

28. I borrowed the phrase "the tragic sense of life" from the title of Miguel de Unamuno's (1972) book, which is actually about death, whose presence makes life tragic. Many thinkers have pointed out that death is the most radical way out of the tragic predicament of life. Here is Nietzsche, in chapter 3 of *Birth of Tragedy*:

> There is an old legend that King Midas for a long time hunted the wise Silenus, the companion of Dionysus, in the forests, without catching him. When Silenus finally fell into the king's hands, the king asked what was the best thing of all for men, the very

> finest. The daemon remained silent, motionless and inflexible, until, compelled by the king, he finally broke out into shrill laughter and said these words, "Suffering creature, born for a day, child of accident and toil, why are you forcing me to say what would give you the greatest pleasure not to hear? The very best thing for you is totally unreachable: not to have been born, not to exist, to be nothing. The second best thing for you, however, is this — to die soon."

For an analysis of the corresponding philosophical stance, anti-natalism, see for instance *Better Never to Have Been: The Harm of Coming into Existence* (Benatar, 2006). Edelman (2020) discusses this and related ideas in chapter 8, *Death*.

29. See (Metzinger, 2017, p.52): "[...] functionally speaking, suffering is necessary for autonomous self-motivation and the emergence of truly intelligent behaviour." Metzinger's proposal concerning the sufferer's options in the face of this evolution-imposed condition has been further discussed by Agarwal and Edelman (2020).

References

Agarwal, A., & Edelman, S. (2020). Functionally effective conscious AI without suffering. *Journal of Artificial Intelligence and Consciousness, 7*, 39–50.

Allefeld, C., Atmanspacher, H., & Wackermann, J. (2009). Mental states as macrostates emerging from brain electrical dynamics. *Chaos, 19*, 015102.

Atmanspacher, H. (2016). On macrostates in complex multi-scale systems. *Entropy, 18*, 426.

Atmanspacher, H. (2017). Contextual emergence in decompositional dual-aspect monism. *Mind & Matter, 15*(1), 111–129.

Baars, B. J. (2005). Global workspace theory of consciousness: Toward a cognitive neuroscience of human experience. *Progress in Brain Research, 150*, 45–53.

Bach, D. R., & Dayan, P. (2017). Algorithms for survival: A comparative perspective on emotions. *Nature Reviews Neuroscience, 18*, 311–319.

Benatar, D. (2006). *Better never to have been: The harm of coming into existence*. Oxford: Oxford Press.

Borges, J. L. (1922/1999). The nothingness of personality. In E. Weinberger (Ed.), *Selected non-fictions* (pp. 3–9). New York, NY: Viking.

Borges, J. L. (1941/1970). Tlön, Uqbar, Orbis Tertius. In D. A. Yates & J. E. Irby (Eds.), *Labyrinths* (pp. 27–43). New York: Penguin.

Cowen, A. S., & Keltner, D. (2017). Self-report captures 27 distinct categories of emotion bridged by continuous gradients. *Psychological and Cognitive Sciences, 114*, E7900–E7909.

Crutchfield, J. P. (1994). The calculi of emergence: Computation, dynamics, and induction. *Physica D, 75*, 11–54.

Crutchfield, J. P., James, R. G., Marzen, S., & Varn, D. P. (2014). Understanding and designing complex systems: Response to ‘A framework for optimal high-level descriptions in science and engineering—preliminary report’. Working Paper 14-12-048, Santa Fe Institute. arxiv.org:1412.8520 [cond-mat.stat-mech].

de Unamuno, M. (1972). *The tragic sense of life in men and nations*. Princeton, NJ: Princeton University Press. Translated by A. Kerrigan, with an Introduction by S. de Madariaga and an Afterword by W. Barrett.

Dehaene, S., King, L. C. J.-R., & Marti, S. (2014). Toward a computational theory of conscious processing. *Current Opinion in Neurobiology, 25*, 76–84.

Edelman, S. (2008). *Computing the mind: How the mind really works*. New York, NY: Oxford University Press.

Edelman, S. (2012). *The happiness of pursuit*. New York, NY: Basic Books.

Edelman, S. (2020). *Life, death, and other inconvenient truths*. Cambridge, MA: MIT Press.

Edelman, S., Moyal, R., & Fekete, T. (2016). To bee or not to bee? *Animal Sentience, 1*, 124. A commentary on *Insects have the capacity for subjective experience*, C. Klein & A. B. Barron, Animal Sentience 2016:100.

Fano, V., Graziani, P., Macrelli, R., & Tarozzi, G. (2016). Are Gandy machines really local? In V. C. Müller (Ed.), *Computing and philosophy: Selected papers from IACAP 2014. Synthese library: Studies in epistemology, logic, methodology, and philosophy of science*, (Chap. 3, vol. 375, pp. 27–44). Cham: Springer.

Frith, C. D., & Metzinger, T. (2016). What’s the use of consciousness? How the stab of conscience made us really conscious. In A. K. Engel, K. J. Friston, & D. Kragic (Eds.), *The pragmatic turn: Toward action-oriented views in cognitive science. Strüngmann forum reports* (vol. 18, pp. 197–224). Cambridge: MIT Press.

Gandy, R. (1980). Church’s thesis and principles for mechanisms. In J. Barwise, H. J. Keisler, & K. Kunen (Eds.), *The Kleene symposium* (pp. 123–148). Dordrecht: North Holland.

Gärdenfors, P. (2020). Events and causal mappings modeled in conceptual spaces. *Frontiers in Psychology, 11*, 630.

Garfield, J. L. (1994). Dependent arising and the emptiness of emptiness: Why did Nāgārjuna start with causation? *Philosophy East and West, 44*, 219–250.

Giudice, G. F., Kehagias, A., & Riotto, A. (2019). The selfish Higgs. *JHEP, 10*, 199.

Harbecke, J., & Atmanspacher, H. (2012). Horizontal and vertical determination of mental and neural states. *Journal of Theoretical and Philosophical Psychology, 32*, 161–179.

James, W. (1890). *The principles of psychology*. New York: Holt. http://psychclassics.yorku.ca/James/Principles/

Kauffman, S. A., & Clayton, P. (2006). On emergence, agency, and organization. *Biology and Philosophy, 21*, 501–521.

Krakauer, D., Bertschinger, N., Olbrich, E., Flack, J. C., & Ay, N. (2020). The information theory of individuality. *Theory in Biosciences, 139*, 209–223.

Llinás, R. R., & Iberall, A. (1977). A global model of neuronal command-control systems. *BioSystems, 8*, 233–235.

MacIver, M. A. (2009). Neuroethology: From morphological computation to planning. In P. Robbins & M. Aydede (Eds.), *The Cambridge handbook of situated cognition* (pp. 480–504). New York, NY: Cambridge University Press.

Marr, D., & Poggio, T. (1977). From understanding computation to understanding neural circuitry. *Neurosciences Research Program Bulletin, 15*, 470–488.

Mayr, E. (1976a). Accident or design: The paradox of evolution. In E. Mayr (Ed.), *Evolution and the diversity of life: Selected essays* (pp. 30–43). Cambridge, MA: Harvard University Press. Originally appeared in Proceedings of the Darwin Centenary Symposium of the Royal Society of Victoria: The Evolution of Living Organisms (1962), 1–14.

Mayr, E. (1976b). Basic concepts of evolutionary biology. In E. Mayr (Ed.), *Evolution and the diversity of life: Selected essays* (pp. 9–16). Cambridge, MA: Harvard University Press. Translated and abridged frorn "Grundgedanken der Evolutionsbiologie," Naturwissenschaften, 56, no. 8 (1969):14–25.

Mei, N., Santana, R., & Soto, R. (2022). Informative neural representations of unseen contents during higher-order processing in human brains and deep artificial networks. *Nature Human Behaviour, 6*, 720–731.

Merker, B. (2012). From probabilities to percepts: A subcortical 'global best estimate buffer' as locus of phenomenal experience. In S. Edelman, T. Fekete, & N. Zach (Eds.), *Being in time: Dynamical models of phenomenal experience* (pp. 37–80). Amsterdam: John Benjamins.

Metzinger, T. (2017). Suffering, the cognitive scotoma. In K. Almqvist & A. Haag (Eds.), *The return of consciousness* (pp. 237–262). Stockholm: Axel and Margaret Ax:son Johnson Foundation.

Minsky, M. (1961). Steps toward artificial intelligence. *Proceedings of the Institute of Radio Engineers, 49*, 8–30.

Morgan, T. H. (1910). Chance or purpose in the origin and evolution of adaptation. *Science, 31*(789), 201–210. Presidential address to the American Society of Naturalists.

Moyal, R., Fekete, T., & Edelman, S. (2020). Dynamical emergence theory (DET): A computational account of phenomenal consciousness. *Minds and Machines, 30*, 1–21.

Ross, D., & Spurrett, D. (2004). What to say to a skeptical metaphysician: A defense manual for cognitive and behavioral scientists. *Behavioral and Brain Sciences, 27*, 603–647.

Rovelli, C. (2020). Agency in physics. https://arxiv.org/abs/2007.05300.

Rovelli, C. (2021). *Helgoland: Making sense of the quantum revolution*. New York, NY: Penguin.

Scharfstein, B. (1998). *A comparative history of world philosophy: From the Upanishads to Kant*. Albany, NY: SUNY Press.

Shalizi, C. R., & Moore, C. (2003). What is a macrostate? arXiv:cond-mat/0303625.

Shanahan, M. (2010). *Embodiment and the inner life*. New York, NY: Oxford University Press.

Simon, H. A. (1973). The organization of complex systems. In H. H. Pattee (Ed.), *Hierarchy theory: The challenge of complex systems* (chap. 1, pp. 1–28). New York: George Braziller.

Smith, J. D., Shields, W. E., & Washburn, D. A. (2003). The comparative psychology of uncertainty monitoring and metacognition. *Behavioral and Brain Sciences, 26*, 317–373.

Smolin, L. (1997). *The life of the cosmos*. New York, NY: Oxford University Press.

Smolin, L. (2018). The dynamics of difference. *Foundations of Physics, 48*, 121–134.

Smolin, L. (2020). On the place of qualia in a relational universe. https://philarchive.org/archive/SMOTPO-3

Sperry, R. W. (1987). Structure and significance of the consciousness revolution. *The Journal of Mind and Behavior, 8*(1), 37–65.

't Hooft, G. (2016). *The cellular automaton interpretation of quantum mechanics*. Cham: Springer.

Wallace, D. (2019). Naturalness and emergence. *The Monist, 102*, 499–524.

Yoshimi, J. (2012). Active internalism and open dynamical systems. *Philosophical Psychology, 25*, 1–24.

3 Self-Consciousness

Self-affirmation. Self-evidencing. Self-creation. Self-location and perspective. Body ownership and agency. Free will. The self as a good regulator. The self's blankie. Subjectivity of self-models. Phenomenal content and phenomenal affect. Virtual reality, real suffering. The third revolution.

> Как известно, наша вселенная находится в чайнике некоего Люй Дунбиня, продающего всякую мелочь на базаре в Чаньани.
> (As everyone knows, our universe is located in the teapot of a certain Lui Dunbin who sells trinkets at the bazaar in Chanyan.)
>
> *USSR Tai Shou Chuan*
> — Victor Pelevin

> Every good regulator of a system must be a model of that system.
>
> *Int. J. Systems Sci. 1(2):89-97 (1970)*
> — R. C. Conant and W. R. Ashby

Self-Affirmation

THE SO-CALLED ontological argument for the existence of God, invented about a thousand years ago by St. Anselm of Canterbury, contends that, as a being no greater than which can be conceived, God necessarily possesses the attribute of existence.[1] In our rather more skeptical age, claims of logically necessary existence of anything, let alone of a supreme being, are no longer seen as convincing. There might be one exception, though: a being's reasoned claim of its own existence does seem necessarily true.

Is an overt declaration of one's own existence — or the mere act of thinking, as per René Descartes' *cogito, ergo sum* argument — indeed proof enough of its truth? Accepting it as such is a simple matter of drawing a logical conclusion from a pragmatic premise: namely, that a nonexistent person cannot think or make declarations.[2] There is, however, another and more potent pragmatic factor here: at a certain level of consciousness (just above that of minimal selfhood, discussed in the previous chapter), one's own existence is an immediate experiential given. This "gut feeling," which informs me that I am, obviates the need for *cogito*-like arguments. There is no need to infer one's existence from one's thinking, as one simply feels it: *sentio ipse, ergo sum.*[3]

S. Edelman, *The Consciousness Revolutions*,
https://doi.org/10.1007/978-3-031-24012-6_3

Where does this feeling come from? Although basic existence is not something that conscious creatures need to monitor, threats to it are. Embodied selves, like you and me, do have a kind of internal pilot light going — an inobtrusive indication that all is well. The elemental experience of being alive and awake is anchored by a low-key, visceral sense of one's own existence — the "internal Sense" of John Locke, "self-feeling" of Alexander Crichton, *Selbstgefühl* of the 18th-century German Philanthropists, and *sentiment de soi-même* of the French Encyclopaedists.[4]

The steady pilot light blinks momentarily when the regulatory mechanisms that monitor the body's physiological well-being detect an imminent danger to its vital functions. The system-wide alarm then becomes the sole and inescapable focus of attention, at the expense of everything else that is being sensed, or planned. The quickest and most acutely felt distress comes from a disruption of breathing, because it runs on such a tight cycle: to maintain an acceptable level of oxygen in the blood, we must inhale and exhale every few seconds. While snorkeling around a reef, you may decide to dive to get a better look at a turtle that is gliding effortlessly in ten meters of water; if you end up holding your breath for too long, the need to get back to the surface becomes the only thing you're conscious of, until normal breathing resumes.

An extinction of the pilot light is not always accompanied by the kind of acute agony that one feels while drowning; nor, indeed, does it always signify mortal danger to the body. Certain malfunctions of the homeostasis-monitoring system may interfere with the feeling of being alive, resulting in strange states of consciousness such as Cotard's delusion, in which people deny their own existence. In this rare condition, the conscious self that is robbed of its inner sense of existence is perfectly positioned to play out the *cogito* argument for all it is worth. The outcome is unequivocal: logic loses to feeling. Cotard patients adamantly insist that they do not exist, all the while being fully aware of their continued cognitive functioning and control over their body.[5]

Self-Evidencing

In being kept on while all is well, rather than saved for when a breakdown needs to be signaled, the inner sense is more like a car's "ready" light than its "check engine" indicator. For whose benefit does it perform, though? In a car, the "ready" light is there to inform the driver; insofar as it is the conscious self that drives deliberate behavior, it needs an inner sense to inform it that all systems are go.[6]

While a system's health can in principle be monitored passively (by watching for signs of possible breakdowns, as done by a car's electronics), the conscious self does it differently. The inner sense signal is thought to be generated by a process of *active inference*, whose task is to seek evidence for the system's continued existence. As noted by Karl Friston, fighting for survival can be seen as an ongoing attempt to maximize such evidence. This is because active search for positive evidence of this kind — but not passive monitoring of the state of affairs — can help the system identify actions, if any, that it should undertake so as best to hold on.[7]

SELF-EVIDENCING is the process of maximizing the evidence for one's own existence, carried out by ACTIVE INFERENCE — that is, seeking the requisite evidence by undertaking actions.

The process of maximizing the evidence for one's own existence, or *self-evidencing*, is future-oriented. In that, it is alike to the faculty of foresight (recall chapter 1), which arises out of dynamic representations of the world being projected into the future. There is, however, a difference. As a computational undertaking, foresight is neutral with respect to its object: the task of predicting where the lightning will strike in a thunderstorm does not a priori distinguish between this location and that; the distribution of electrical charge over the terrain is the only thing that matters. In contrast, self-evidencing is neither impartial nor impersonal: it is disproportionately focused on those possible futures that include the self's existence. Exercising self-evidencing in a lightning storm involves making a distinction between the location of the self — right now, as well as in the immediate future — and all other locations.[8]

As all statistical inference, self-evidencing deals in probabilities. What does it mean? In chapter 1, I described mind-like computational systems as those that maintain variables representing various aspects of the world. In probabilistic representations, the value of a variable may be uncertain, being distributed over a range rather than fixed at a specific number. Because the knowledge that any representational system might have about the world is necessarily incomplete and may be noisy, uncertainty management, in the form of probabilistic learning and inference, is a must.

The kind of inference required for self-evidencing involves a search for conditions that would maximize the probability of the "good" possible futures — those in which *I* exist (in the states that characterize me). The outcome of this computation can then be used to plan actions aimed at bringing about the desired conditions. I have reflexively put scare quotes around "good" just now, but self-evidencing happens to be one of those rare contexts in which such caution in reducing philosophically contested concepts to plain science is unnecessary. It is difficult to imagine a more effective basis for defining what counts as good for an organism than the continuation of its existence (suffering and all).[9]

To be relevant to consciousness, this way of distinguishing between good (i.e., characteristic of me, ego-syntonic, familiar, *et cetera*) and evil (uncharacteristic of me, ego-dystonic, surprising, *et cetera*) must be intrinsic to the system in question. In naturally evolved systems (as well as those that are engineered with lessons from nature in mind), the principle that ensures the intrinsic relevance of goodness being rooted in self-evidencing is evolutionary selection for success. This principle is expressed by the slogan "nothing succeeds like success" from p.39; in the present case, a system that acts as if its continued existence is what's good for it is likely to be able to stave off its demise for longer than one that doesn't.

An intimation of this take on good and evil appeared already in chapter 1 (p.18), where caring was defined in terms of the distinction between outcomes that are conducive to the evaluating system's well-being and those that are not. We have come full circle to that same idea, newly armed with the concept of self-evidencing through active inference. If caring amounts to being intrinsically and inexorably biased towards certain outcomes, then the inner sense of a conscious self is a means of focusing caring on what matters — to the self. Paraphrasing yet again Descartes's dictum, *auguror ipse, ergo sum*: I predict myself, therefore I am.

Self-Creation

A system that is sustained by self-evidencing is a stranger kind of computational entity than anything that we have encountered so far: one that continues to exist by virtue of predicting that it will. Why should this approach to survival be any more effective than wishful thinking? Indeed it shouldn't — unless predictions are accompanied by actions aimed at good outcomes.

On the scale of cognitive-computational sophistication, active pursuit of preferred outcomes is one rung above the reinforcement learning that keeps together the effective or minimal self's "persistent local causal neighborhood" (chapter 2, p.39). The minimal self is the bundle of control processes that orchestrate the *homeostasis*, or tendency toward equilibrium, of the parameters of the system that computes it. By developing the capacity for self-evidencing, the now no-longer-minimal self can also do *allostasis*, or active and anticipatory (rather than merely reactive) control.

The concept of homeostasis is familiar to engineers who design automatic control systems. A simple example of a homeostatic system is a room with thermostat-controlled air conditioning. The A/C unit is switched off, or set to cooling or heating, if the temperature is within specified limits, too high, or too low. The possibility of achieving homeostasis through *self*-regulation explains why it crops up in a computational account of (minimal) selfhood.[10]

Homeostasis is of course also familiar to biologists (and philosophers of biology) who study the minimal conditions for a system to qualify as being alive. A widely agreed-upon characteristic of living systems is their need to maintain certain vital parameters within a specific range — and to do so on their own. There is a deep analogy between life-like and mind-like processes: on the one hand, the physiological adaptability and self-regulation that make life possible; on the other hand, the computational learning and self-evidencing that make some life forms minimally conscious.[11]

There is an analogy between life-like and mind-like processes: on the one hand, the physiological adaptability and self-regulation that make life possible; on the other hand, the computational learning and self-evidencing that make some life forms minimally conscious.

Just like consciousness is necessarily intrinsic to the system in question, so is physiological homeostasis. A living cell keeps itself alive; and a multicellular organism such as a human being does not, in its normal state, require external life support. In theoretical biology, intrinsic and autonomous maintenance of homeostasis is sometimes referred to as *autopoiesis* or self-creation.[12] No miracles here, just plain thermodynamics: an autopoietic system performs a work cycle, comprising a constrained release of energy, which is applied to construct and maintain those very constraints. The same kind of cycle — albeit informational, not metabolic — underlies the operation of a computational system (no matter how it is implemented) that is engaged in self-evidencing: information is acquired under the guidance of predictive processes and is then used both to ensure that the mechanisms that implement the computations survive and to improve their predictive power. The emergent virtual entity that wields that power is a personification of the prediction function, whose unity and global availability makes it effective at the task of keeping the system together and easily invoked by any of its components. Quoting Rodolfo Llinás,

> What is the repository of predictive function? The answer lies in what we call the self, i.e. the self is the centralization of the predictive imperative.[13]

We should keep in mind that there is a long way from this level of selfhood to full-blown human consciousness. In the just-quoted passage, Llinás proceeds to note that "The self is not born out of the realm of consciousness — only the noticing of it is (i.e. self-awareness). Thus, according to this view, the self can exist without awareness of its existence. That is, even in our case of individuals capable of self-awareness, such awareness is not necessarily present." Quite a few functional components need to be added to the "center-of-prediction" construct to make it conscious in the same way that we are. The first of these is the visual sense of self-location and of perspective onto the visible world.

Self-Location and Perspective

Open your eyes, and you will have the experience of occupying a definite location in space: the place where you are right now, from which you regard the visible part of your world. This mode of self-location depends on the ability to see, which in turn depends in part on the evolutionary history and the behavioral needs of the species in question. The self-location needs of a bacterium are modest: for an *E. coli*, it is useful to know whether or not it is inside a host's gut and in which direction the concentration of nutrients is higher. A squirrel, in comparison, needs to know much more: the spatial arrangement of accessible tree trunks and branches, potential food sources, and signs of danger — and its own location with respect to all these.

Our needs are much like the squirrel's; indeed, we share the major characteristics of our conscious visual experience, including self-location, with all vertebrates and many insect species. Our experience of self-location, sometimes referred to as *perspectivalness*, is notably easy to subvert: wearing a display connected to a camera teleports the viewer to the camera's vantage point, in a manner that is especially convincing if the camera is yoked to the viewer's head movements.

Interestingly enough, the full power of the brain to instill a sense of self-location and perspective is revealed when visual information is fed to it indirectly, for instance through the skin. In such a *sensory substitution* setup, a camera is connected to an array of small pins that are pressed against the small of the subject's back. Each pin is made to vibrate with an amplitude that is proportional to the light intensity of the corresponding pixel in the camera's output. Very quickly, the subject (who may be blind) learns to see, through the skin of their back (thus at a rather low resolution), the scene that the camera is pointing at. Thereafter, all the tricks that can be played on the brain by manipulating the sense of sight, such as teleporting the viewer to another location, become available through the sense of touch, coopted to serve vision.[14]

The ease with which one can see through one's skin has a simple and straightforward explanation. As noted back in chapter 1 (p.9), our impression of having immediate visual access to the real world, seen through a hole in the front of the skull, is an illusion: the world we see is a virtual construct and there is nothing immediate about the process whereby the brain computes it. When this process is fed information that can be interpreted in the same manner as the retinal image is, it proceeds with constructing the virtual world from that information.[15] Moreover, it does so while hiding from the (equally virtual) self that inhabits this world both the source of the signal and the workings of the processing pipeline. This last part is of course exactly what happens in "normal" vision, too: what we see is neither the image formed by the eye's lens on the retina, nor the neural representations of that image, but rather the world.

The really amazing thing about how the brain computes the visual world is how open this computation is to sources of information other than the eyes. Seeing through the skin is a beautiful illustration of this openness, but there's plenty of evidence for it in the routine operation of our sense of sight. Look up from this page for a moment; did your visual world swing down as your eyes moved up? Probably not. There is a special process in the brain that keeps the visual world stable, despite the fast and abrupt eye movements, called saccades, that happen on the average three times a second as the eyes fixate various points of interest.

This process is very economical: instead of trying to undo the consequences of the saccades, the visual processing pipeline is instructed to simply ignore them. Every time an eye movement is commanded, an *efference copy* of the command serves as the signal to temporarily disconnect the pipeline, to shut out the momentarily blurry input; and to disregard the imminent shift of the retinal image, so that the virtual world remains phenomenally stable, even as the retinal projection of the real world swings this way or that.[16]

The standard vertebrate body plan (which we share with squirrels) allows for independent movement not just of the eyes, but also of the head, which is where our eyes are housed. Given how many muscles it takes to control head orientation, compensating for head movements through the use of an efference copy would be too unwieldy, which is probably why an actual measurement of head attitude, provided by the vestibular sense, is used instead. Reliance on this little shortcut has spectacular consequences when parts of the vestibular system fail: a malfunction of the utricle and saccule, the two inner-ear organs that signal head tilt, causes the visual environment to be perceived as tilted.

Patients may experience this "room tilt illusion" in different ways: for instance, the room may appear tilted sideways to various degrees, or the ceiling — complete with the light fixture dangling from it at a right angle — may take the place of the front wall.[17] The illusion is so irresistible that no amount of reasoning from the physical implausibility of the tilted environment (mugs miraculously staying put on the no-longer horizontal tabletop, etc.) can dispel it. It serves as another reminder of a basic fact of vision, mentioned a few paragraphs back: we *don't* see the retinal image as such. After all, the retinal image is not different from normal in people in whom vestibular malfunction causes the room to be seen as tilted.

What is the retinal image like, normally, though? The eye's lens is doubly convex, so the image of the scene it projects onto the retina must be upside down. This hypothesis was verified in the 1630s by René Descartes, who scraped the flesh off the back of an ox eye, until he could see through the remaining thin tissue the inverted image formed by the lens. How come we don't see the world upside down, then? Descartes was also the first to offer an essentially correct explanation of why there is no need to re-invert the retinal image anywhere in the brain: "c'est l'âme qui voit, et non pas l'œil, et elle ne voit immédiatement que par l'entremise du cerveau" (it is the soul that sees and not the eye, and it immediately sees through the intervention of the brain).[18]

The self computed by the brain has no intrinsic reference to rely upon in deciding which direction is up. Neither is absolute reference available in the visual information obtained from the outside world. Certain consistent patterns in this information, such as the motion of objects in free fall, suggest that not all directions are equivalent — but that still does not mark the retinal image in itself as either right side up or upside down. Consequently, looking at the visual world exclusively through image-inverting eyes, the self naturally experiences its orientation as normal.[19]

Body Ownership and Agency

The contribution of vision to the human phenomenal experience is so weighty that it is capable of overriding a sense that would otherwise seem to be fundamental and immune to interference: that of one's own body. A most striking example of such an override — and a proof that the body (part) *ownership* aspect of the self is, like all its other aspects, virtual — is the rubber hand illusion, first reported in 1998.[20]

To experience this illusion, sit at a table, with one hand on the table and the other on your knee, out of sight. Place a rubber hand on the table in front of you, on the side of your invisible hand. Have another person use a pencil to stroke back and forth the rubber hand and, simultaneously, your real hand that's under the table. You will find yourself experiencing the rubber hand as yours.

This illusion often works even with objects that don't look like a hand at all — you can try it with a small oblong box or a book. (Not that this version of the illusion is more surprising than the original; after all, why would an object that looks like a hand feel like it's your hand?) What is critically important is that the vision and touch cues be properly coordinated: breaking the synchrony of stroking the visible object and the hidden hand disrupts the illusion.

These observations reveal something about the computations that maintain the normal sense of body (part) ownership — and that allow it to be subverted. The key explanatory concept is that, which the self is the center of: prediction. The first step towards making it possible to predict goings on in the world — of which the body is part — is to maintain a tally of what cues co-occur or, more formally, to estimate the joint statistics over the relevant variables. The resulting table of co-occurrence counts can then be used to "predict" one cue or event on the basis of another by computing the conditional probability of it happening, given that the other has been observed.

The tally table is a kind of a model of the world — a simple one, as models go, and easy to learn (easier than the dynamical models of various world processes, first mentioned in chapter 1, p.17). Its predictive power is, however, limited, because joint statistics capture correlational information only. To learn about the causal structure of the world, which is what a dynamical model attempts to do, additional work is needed (more about this later in this chapter). Still, the knowledge of prevailing correlations can lead the self to expect one cue, given another. Returning to the question of hand ownership, the expectation based on a lifetime's worth of statistics is that if I see an object being stroked at the same time as the skin of my hand is felt to be, then the object must be my hand.[21]

The self's reliance on correlation-based predictions to figure out its body ownership, which is co-opted in the rubber hand illusion setup, is a very general computational principle. The loopholes that it suffers from have recently been exploited to induce subjects to feel that they see their body from another location in the room (an out of body experience or heautoscopy); that their body is doll-sized or giant; that they simultaneously have two bodies; that they occupy two distinct locations; and more.[22]

A tacit condition imposed on the subjects in all the experiments patterned on the rubber hand illusion is that no movement of the body part(s) in question be initiated. Obviously, even as the feeling of ownership over a rubber hand is induced in me by properly synchronized visual and tactile stimulation, any attempt on my part to move "my" hand would immediately dispel the illusion. Importantly, the concept of prediction still applies: I command a hand to move; because it is my hand, I predict that it will move as commanded; it fails to move; thus, it cannot be mine.

A complementary failure of prediction occurs when a body part moves without the move being willed. A vivid fictional example is the unruly hand in Stanley Kubrik's 1964 film, *Dr Strangelove*, which in real life would be classified as a case of the alien hand syndrome.[23] According to the principle of active inference, which underlies the inner sense and the process of self-evidencing discussed in the beginning of the chapter, a movement is caused by first predicting the desired outcome, then acting to make it come to pass. If the action has not been willed or commanded, the discrepancy between the predicted lack of movement and the factual outcome is experienced as a loss of *agency*.

A minimal sense of agency exists in all systems that can act on their own behalf and that must therefore distinguish between what they do and what is done to them.[24] This functional aspect of minimal agency makes it part of the minimal effective self, which, as we saw, is found, among living systems, already at the level of single-cell organisms. In humans, the cluster of feelings having to do with agency is much more complex, but its core is still the sense that grows out of the predictive function at the center of the multifaceted and sophisticated self.[25]

The computational foundation of this predictive function is a familiar one. In animals like us, the sense of body ownership and the sense of agency both rely on the use of efference copy — the same principle that helps the brain keep the virtual visual world it constructs steady, despite constant eye movements. In any self-initiated movement, a planning process first maps an intended state to a motor command, which is then sent out for execution, with a copy being made available for internal reference. Body ownership can subsequently be inferred from a match between (i) the predicted state computed from the efference copy and (ii) sensory feedback that follows the execution of the command. Similarly, agency can be inferred from a match between (i) the intended state and (ii) the predicted state.[26] But where do intended states come from?

Free Will

In our daily experience, questions about the origins of our intentions — say, "Where did this urge to write a section on free will come from?" — do not arise. When asked, such questions seem trivial: the urge came from me, where else? I willed it freely. Barring rare disorders, such as the delusion of thought insertion, people *feel* themselves to be free agents. Complications arise only when we pause to think about this feeling of freedom: Where does it spring from? Is it factually justified? Does it even make sense?

This last question, which taps into the philosophy of free will, is remarkably easy to settle, given how contentious it still is among some philosophers and how confusing the popular science debates that surround it can be. The answer to this question comes in two parts, one negative and the other positive:

(I) In any matter in which the would-be agent's choice is not influenced by some prior event(s), it must necessarily be down to an *uncaused cause*. In each such matter, the agent effectively ignores the rest of the universe, and its absolute freedom comes at the expense of being "not of this world." *Corollary:* Making decisions on the basis of random choice[27] disrupts the agent's engagement with the world, as indicated.

(II) In any matter in which the prior events that influence the choice at hand have been incorporated into the would-be agent (as memories, or as predispositions, either built in or acquired), its exercise of agency is real and free, in a sense that matters.

To see why part I holds, recall the discussion in chapter 2 of the universe-spanning web of cause and effect. Within that great web of causal relationships, effective selves emerge as self-consistent and self-perpetuating local causal domains. Critically, an emergent self remains causally coupled to its environment (as any part of the universe must). This implies that all of the self's internal processes are potentially open to environmental inputs. These "boundary conditions" constrain the computations that constitute the self, so that none of its choices or actions can be absolutely and unconditionally free, that is, free of all causal influences from the outside. A system that is free in this absolute sense — by means of implementing an uncaused cause — is half-disengaged from the world and cannot appear to it except as a source of random interference.[28]

To better understand how outside influences shape people's decision-making and their fate, consider the following true story, recounted by the great Polish writer, Stanisław Lem, in an autobiographical essay:

> That chance played a role in my life is undeniable. In the First World War, when the fortress of Przemysl fell, in 1915, my father, Samuel Lem, a physician in the Austro-Hungarian Army, was taken prisoner by the Russians, and was able to return to Lemberg (now Lvov), his native city, only after nearly five years, in the wake of the chaos of the Russian Revolution. I know from the stories he told us that on at least one occasion he was to be shot by the Reds on the spot for being an officer (and therefore a class enemy). He owed his life to the fact that when he was being led to his execution in a small Ukrainian city he was noticed and recognized from the sidewalk by a Jewish barber from Lemberg who used to shave the military commander in that city and for this reason had free access to him. The barber interceded for my father (who was then not yet my father), and he was allowed to go free, and was able to return to Lemberg and to his fiancee.[29]

Whatever other reasons I may have had for including this anecdote here, my decision to do so depended critically on one cause that is remote both in space and in time: the barber's intercession on behalf of Lem's future father. Had the barber of Lemberg stepped out of his shop a minute later, missing his chance to save the condemned man, Lem would not have been born, and I would have to come up with another story to tell. More likely, though, there would be no need for any story: without Lem's books, which I read during my formative years, I would not be writing this one (and you would not be reading this, which means that the barber is in your causal past, too).[30]

Lem calls the unlikely cause behind his father's narrow escape "chance." As in most such cases, however, this label merely reflects an observer's inability to trace all the causal influences behind the event in question. This inability typically stems not so much from the limitations of the observer as from the objective computational intractability of prediction in a complex system. In particular, a completely deterministic system, such as thrown dice, can exhibit chaotic behavior, in which infinitesimally small changes in the initial conditions rapidly cause a wide divergence in the outcomes, making the end result effectively unpredictable. True randomness, in contrast, is rather rare in nature: it characterizes, for instance, radioactive decay.[31]

In any case, unpredictability of outcomes, whether it stems from deterministic chaos or from true randomness, is not freedom. Suppose you wish to claim agency in some matter of interest, in which you face a choice. If you predicate your decision on the decay of an atom of radioactive cesium ^{137}Cs (or on the fall of a perfect die), you are not exercising any kind of freedom: rather, you cede the choice, and with it your agency, to chance.

Let us now turn to part II of the answer to the question of free will. This part focuses on cases in which a trace of the events that affect the present situation has been incorporated into the agent. Was the barber from Lem's story an agent in his own right, or was he merely a conduit for the effects of earlier events that culminated in him saving Lem's father — stretching perhaps all the way back to the Big Bang?

Agency, as it is explained here, can be given a formal definition in the language of probability theory. A system's state is causal — making the system effectively an agent — when it renders the system's future behavior *conditionally independent* of its past.
— James Crutchfield (1994, p.26).[32]

Those of us who have been hoping all along that our experience of being free agents is grounded in reality will be pleased to learn that it is. Even if the Big Bang is *the* ultimate cause of everything that has been happening in this universe, the proximate — and entirely sufficient — causes for an agent's choices reside within the local causal domain that constitutes its effective self.

A general label by which these agent-internal causes go is *memory*. Thus, in Lem's story, prior acquaintance with Lem Sr. had effected a lasting change in the barber's brain, in the form of a memory record. Insofar as that memory was part of the barber's self, its activation upon visual exposure to Lem Sr. was an event internal to the barber, which, in turn, caused the barber to intervene. This demonstrates that there is no need to pass the buck of causation on beyond the barber and back to the beginning of time: the barber *was* the agent of his action.

Was the barber *conscious* of his agency, though? Probably not. Our routine voluntary acts, which arise out of a churning cauldron of subconscious computations involving memory, perception, and thinking, are "marked by the absence of surprise" — an insightful characterization of agentic action offered by Ludwig Wittgenstein.[33] It is worth noting that Wittgenstein's phrasing is somewhat self-contradictory: voluntary action is *not* marked in any special way, precisely because as a matter of routine there is no surprise.

Now if my arm were to go up without me willing it, that would be surprising. Such an occurrence would preempt everything else that I might be doing at the time and would likely take over the entirety of my consciousness. The same negative characterization applies to one's sense of embodiment. Normally, I am not conscious of the positions of my limbs or of the fabric of my clothes touching my skin. Only when something goes wrong in the body sense department is the conscious self alerted (by a disturbance in the inner sense) and recruited to the effort of righting the wrong. Thinking, too, is experienced like that. It does not feel like anything is out of order when in my head ideas spring up, thoughts chase one another, and choices emerge. Thoughts can, however, feel intrusive: as being merely "out of the blue" or, in pathological cases, as having been implanted and as such not belonging to the self.[34]

It is just as well that the phenomenal experience of being a conscious, embodied, and thinking agent is sparse, in contrast to how rich it appears to philosophical deliberations.[35] Its contents are picked out from the ongoing massive flow of internal and external stimulation by the "flights and perchings" of attention, without which the cognitive system would be overwhelmed. This stream-of-consciousness spectacle in the head, for which the self serves as the producer, the director, the animator, and the spectator, plays out against a constant low-level background hum of inner sense, whose function is to reassure the crew that all is well.[36]

Who is in charge of this show? A virtual persona, arising, if need be, out of the interactions among the many subconscious processes that make up the production team, all of which are sub-personal and very real. The virtual conscious self, the "center of prediction," consolidates the control of behavior, making it more effective when the interests of the system as a whole are at stake.

While the freedom of the show's virtual director is checked by the system's past and present interactions with the rest of the world, its illusion of being in control is complete. We all feel free and unconstrained in our choices, as if the laws of physics do not apply to our brains. This experience of conscious will has been called "the mind's best trick."[37] The computational reason for the existence of this illusion, which has of course been shaped by evolution, arises from the vital need to model the modeler as part of the world — all in the service of the true categorical imperative: prediction.

The Self as a Good Regulator

We have seen already in chapter 1 that it is in the agent's best interest to represent the dynamics of its world, so as to be able to predict which way things are headed and to plan accordingly. The argument for prediction can actually be made into a formal theorem. The work of Suzanne Still and others has shown that the need to be predictive stems from basic thermodynamics: "any system constructed to keep memory about its environment and to operate with maximal energetic efficiency has to be predictive."[38]

The thermodynamic formulation does not specify how the capacity for prediction should be incorporated into behavioral control. For that, we need to look at the theory of reinforcement learning (which also already came up in chapter 1). The objective of reinforcement learning is to maximize the cumulative reward that the agent gains from its ongoing interactions with the world. There are two ways of going about this. One approach is to try and learn something about how certain things in the world work, build a model that captures this knowledge, and leverage it to select the best actions in each situation. The other skips explicit modeling of the processes that drive world events in favor of tabulating associations among circumstances, actions, and rewards.[39]

It is the first, model-based style of reinforcement learning that points the way to the emergence of the self. The model that is being learned is an internal computational representation of the world, or rather the part of it that is relevant to the task at hand. Its goal is explanatory rather than merely descriptive: to account for the observable aspects of the world in terms of states that are not directly observable. These "hidden" states, and the transitions between them, are inferred by the learning algorithm from the observed data. The resulting knowledge of how things work allows the agent to exercise forethought, by generating probabilistic predictions of outcomes for each contemplated course of action, given an informed hypothesis about the present state of the world.[40]

In an evolutionary setting, the scope and the sophistication of the models vary greatly across species. An *E. coli* bacterium may do very well inferring and acting on just a couple of variables that characterize the immediate environment in its corner of the gut. In comparison, humans routinely monitor and use, and sometimes explicitly reason about, the hidden causes of many things (including, *inter alia*, what the microscopic commensals in their gut are up to).

Clearly, how simple or how elaborate a particular species' models get to be depends on the ecological niche in which the modelers reside. In any case, the ubiquity, in naturally evolved systems, of modeling the situation outside oneself suggests that having a model always beats not having one. There are two distinct formal arguments that can be brought to bear on this hunch. The first one shows that having a model of a system is indeed obligatory for exerting optimal control over it; the second one (discussed in the next section) spells out *why* having such a model is the controller's most vital interest.

The first and basic argument in favor of investing in a model draws on a theorem, proved in 1970 by Roger Conant and Ross Ashby, which declares that "every good regulator of a system must be a model of that system."[41] The theorem applies to a setup that consists of a target system and a controller, each of which is defined by the set of states they can be in. Both the system's and the controller's states may bring about outcomes, some of which are deemed "good" or desirable. Both the system and the controller are affected by inputs that perturb their states and thus influence outcomes. In Conant and Ashby's formulation, the controller's "regulation" of the system consists in acting alongside it, and its "goodness" is defined as the ability to keep the outcomes ordered, in the sense of being stable or relatively constant over time (as quantified by their entropy, which is a measure of disorder).[42] An example of this kind of good regulator is the brain process that homeostatically controls our body temperature, so that perturbations in the form of overheating or hypothermia are counteracted, respectively, by a command to increase sweating or to shiver.

[T]he probability of [a living system's] (interoceptive and exteroceptive) sensory states must have low entropy; in other words, there is a high probability that a system will be in any of a small number of states, and a low probability that it will be in the remaining states. Entropy is also the average self information or 'surprise' (more formally, it is the negative log-probability of an outcome). Here, 'a fish out of water' would be in a surprising state (both emotionally and mathematically).
— Karl Friston (2010, p.127)

The Self's Blankie

In setting the stage for the Good Regulator theorem, Conant and Ashby had introduced notation that distinguished between good outcomes and others, but then proceeded to define good regulation as the pursuit of outcomes that are merely stable, or low-entropy (in that the changes in the outcomes produced by the regulator's responses to the system are minimal). It would seem that this definition gives up on the search for good outcomes, but it turns out that, in the regulation of behavior in the face of disorder and disruption, "stable" can actually mean "good."

The concept that links low entropy with desirability is homeostasis. In a world where dangers abound and even the very passage of time spells decay, all living systems must actively stabilize the values of their vital variables within a certain range. This implies that stability as such is necessary for survival. More importantly, keeping the vitals within the proper range is not only necessary but also sufficient for staying alive. Evolution ("nothing succeeds like success") can thus be expected to lead to the emergence of systems for which homeostasis is inherently desirable — in other words, good (without qualifications or scare quotes). Good regulation is the cornerstone of life.

In living systems, the range of tolerable values of vital parameters is typically narrow compared to their full possible range (think about how tightly our body regulates its temperature or the concentration of oxygen in the blood). This means that a homeostat imposes a high degree of order on the system it regulates — that is, low uncertainty, or, in the language of thermodynamics, low entropy (disorder or high uncertainty/entropy would correspond to the values spread randomly over a wide range). But low uncertainty just means low *surprise*. This suggests a characterization, both mathematically sound and intuitively appealing, of what it might mean for an outcome to be inherently good for a living system, and thus to serve as a worthy goal for a good regulator: one that contains no surprises.

An outcome is inherently GOOD for a living system insofar as it contains no surprises.

It is a truism that to avoid surprises in life one must be endowed with the faculty of foresight — that is, an ability to anticipate impending events. This, in turn, requires an active kind of understanding of how things work: one that employs a model that can generate predictions, including explicit estimates of the likely effects of one's intended actions. The mathematical theory behind such *generative models* integrates classical results in probability, going back to a 1760s theorem due to Thomas Bayes, with recent conceptual insights into the computational nature of prediction and surprise, based on the work of Karl Friston and others.

A straightforward attempt to use the minimization of surprise in guiding the regulation of behavior runs headlong into a problem: to figure out just how surprising (or not) a state of affairs is, one must know the circumstances that led to it, which in principle may range over the entire universe. Should I be surprised by the series of strikingly red sunsets, the likes of which are rarely seen in this corner of upstate New York? Not if I know that out west, where our weather comes from, the wildfires of California, intensified by global warming, have for weeks been pouring smoke into the troposphere.

One cannot expect such knowledge to be easy to come by. In the days before instant transcontinental communication, the sunset question would have been moot — but the very same kind of difficulty arises, on a smaller scale, in the daily life of any creature that must engage its world, juggling perception, inference, and action for survival. Why is that frog sporting such a vivid neon color scheme? Perhaps I should not try to catch and eat it, after all. Why is there a buzzing sound coming from that desert bush, with no dry mesquite pods within sight? Must be a rattlesnake there, trying to help me not be surprised by it.

In fact, the surprise-minimizing agent's predicament is more dire than that. By dint of its necessarily limited spatial extent, the agent — embodied in a local causal domain — does not have access to knowledge about what is going on "out there," except through a set of variables that form its interface with the rest of the world. This set of variables, whose states are directly accessible to the agent (and, from the outside, to the world), is called the Markov blanket, a concept which I first mentioned in chapter 1 and which has been named after a 19th century mathematician who made key contributions to probability theory.[43] In the case of my brain, the blanket consists of those neurons that track incoming sensory inputs, as well as those that send outgoing signals to the muscles. Focusing for the moment on the former, it should be clear that the states that are directly observable by my brain — the activities of the photoreceptors in my retina, or of the cochlear hair cells in my inner ear — are the informational pivot on which my assessment of surprise, as well as all my other inferences about the world, must necessarily hinge.

Still, the blanket-wrapped agent has a recourse: what it cannot *know* about the world, it can try to *model*, using the information that is available to it at the blanket. Learning and maintaining an internal model of the hidden external causes of the blanket states creates an opening for the estimation of surprise (which otherwise would be an impossible undertaking). The resulting estimate takes the form of an upper bound: a quantity that is mathematically guaranteed to be larger than the unknowable true value of surprise. By driving this quantity down, the agent can thus ensure that surprise too is minimized.

An information-theoretic analysis of surprise reveals that this upper bound, called *free energy*, can be completely described in terms of two probabilistic quantities, whose estimation in turn requires only information that is available to the agent: (i) its sensory states, that is, the blanket states that it can access directly; and (ii) an educated guess about the external causes of these states, generated by its internal model of the world. Because true surprise is always smaller than free energy, working to minimize either of these two quantities allows the agent effectively to minimize surprise.[44]

Each of the two components of free energy offers a distinct opportunity for minimizing surprise. It's easier to start with the second component, which can be manipulated by improving the internal model of the world (presumably, through reinforcement learning): the better the model, the more probable the actual observations, and the more precise the explanations of these observations in terms of their external causes. This is the essence of *perception* — not merely of the blanket states, but of the otherwise "invisible" reality behind the veil. Thus, when I look at the world out of my eyes, what I see is not the light patterns on my retina, but rather the objects "out there," whose presence and appearance caused the cascade of neural activity in my brain.

What about the first component of free energy, the sensory states? It would seem that the agent has no control over these: what it sees and hears is determined by the external world. Not so: an embodied agent can typically *act* so as to make a particular perceptual state more probable. A hungry *E. coli* bacterium, presently residing in my intestine, can swim in the direction of a higher concentration of nutrients, to help come true its prediction of a better place to be, for a while; and the hungry me can get up and go over to the fridge, to fulfill my prediction of soon facing lunch instead of this screen.[45]

It is important to remember that such predictions are only as good as the model that generates them. This does not imply that a bacterium and a human both need the most sophisticated model that can be had — merely that the model must be good enough to support the range of behaviors that is typical for the species to which its owner belongs.[46] And so it goes: a good model means good predictive power, which means less surprise. The conceptual connection here runs deep: improving the quality of the model, or *model evidence*, which must be done to minimize surprise, is mathematically the same goal as self-evidencing — a notion that came up earlier in this chapter, in the discussion of the inner-sense component of consciousness.[47]

Subjectivity of Self-Models

The upshot of the above two-pronged argument for the agent's need to model the world, based on the Good Regulator theorem and the minimization of surprise, is that the controller's dynamics — states and their transitions in response to any inputs — must mirror the dynamics of the controlled system. Because the dynamics of the brain is what phenomenal experience consists in, the Good Regulator theorem adds a key ingredient to the understanding, which has been with us since chapter 1, that the mind inhabits a virtual reality constructed by the brain. Clearly, the virtual reality that we experience must be reliably related to the real world "out there," or else we would not have survived the evolutionary test of time.[48] Mere reliability is, however, a rather vague constraint, which can be met in many different ways. The Good Regulator theorem provides a more concrete constraint by suggesting that our phenomenal experience derives from running an internal model of the world.

> PHENOMENAL EXPERIENCE
> ... is the system's trajectory through its set of macrostates.
> —Moyal et al. (2020, p.6)

Given that the self, too, is a virtual construct, we may ask: what does it take for a controller to count as a self, and what, if anything, is *that* a model of? So far in this chapter we saw that the experiential components of this construct include: a sense of continued existence; a perspective on the world; body ownership; and agency. All of these qualities are representational, and each of them stands for something real: for being alive, awake, and aware; for being in a specific location; for occupying a body; for having a say in that body's behavior. Collectively, the processes that regulate these qualities, which together amount to self-consciousness, do serve as a controller, which, moreover, happens to be a model. It is the *phenomenal self-model* — the controller's model of itself. As we shall see, it is the peculiar self-referential nature of systems that contain their own models that makes them capable of subjective experience.[49]

Thomas Metzinger, who formulated the concept of phenomenal self-model that I just described, has likened a system that is capable of maintaining such a model to a *total flight simulator*.[50] A regular flight simulator, used in training pilots and for gaming, consists of a computational model that simulates the airplane's behavior in response to the pilot's commands and the terrain over which the plane flies. Metzinger's concept extends this idea: in a total flight simulator, the pilot is simulated along with the plane and the terrain. What controls the plane, then, is a virtual construct — a model of the pilot — which becomes self-conscious by virtue of simulating itself as such.

It is somewhat reassuring to know that this dizzying stunt on part of an entity that is virtual to begin with has roots in the mathematics of control. The need for the agent/controller to simulate itself along with the controlled system ("the world") is a simple consequence of the Good Regulator theorem. The basic premise is that the agent is dynamically embedded in the world: if it were not, there would be no way for it to control anything. In acting so as to regulate the world, the agent therefore necessarily also acts on itself. This would seem to imply that it must include in its model of the world a model of itself.[51]

A control scheme in which the controller's interventions on the world reach back to affect its own state is very familiar to engineers: it is called a feedback loop. The mathematics of feedback control is well understood and is widely relied upon in the design of applied control systems, from a home thermostat to an airplane autopilot. There is nothing mysterious or conceptually problematic about feedback control, not even when the controller is made to rely on a model of the system it controls. But if we follow the Good Regulator theorem through and set out to build a controller that models the world as well as itself, a certain complication does arise.

Some symbolic notation will help us see the nature of this complication. Let us denote by m the modeling function that is employed by the controller/self $\mathscr{S}$ — a dynamical system whose purpose is to regulate another dynamical system, the world $\mathscr{W}$ (as per the Good Regulator theorem). If "the world" is taken to consist of the environment $\mathscr{E}$ and the self taken together, expressed formally as the product $\mathscr{W} = \mathscr{E} \times \mathscr{S}$,[52] then we can write

$$\begin{aligned}\mathscr{S} &= m(\mathscr{W}) \\ &= m(\mathscr{E} \times \mathscr{S}) \\ &= m(\mathscr{E} \times m(\mathscr{E} \times \mathscr{S})) \\ &\ldots\end{aligned}$$

You can see the complication now: our entity of interest, the self, is invoked inside its own definition, which makes it *recursive* and leads to infinite regress if one attempts to "spell it out." Because the definition of the self (the second line in the above equation) is self-referential, substituting it for the symbol $\mathscr{S}$ leaves an expression (the third line) that still contains $\mathscr{S}$. No matter how many successive rounds of substitution we attempt, the resulting formula, growing longer with each round, would still contain $\mathscr{S}$, thus requiring further expansion, *ad infinitum.*

Of course, notation is not the problem here: the nature of this kind of self is. Any controller that evolves or is engineered so as to apply the lesson of the Good Regulator theorem *and*, moreover, to include itself in its model of "the world" will risk getting bogged down in recursion the moment it calls upon the model that defines it, for purposes of forethought and action planning. Arguably, the perspective of sliding into infinite regress is the source of the sense of dizziness that we experience when we turn our mind's eye inward in an attempt to pin down the mind's *I*.[53]

The existence of a self-referential process or "strange loop" at the core of the self has been noted and discussed by many scholars and theorists of consciousness. It has been given a brilliant book-length treatment by Douglas Hofstadter (the author of *Gödel, Escher, Bach* and *I Am a Strange Loop*), who has discussed the manifestations of self-reference in our sense of self in the context of the computational concept of recursion. There is also much evidence from cognitive and computational neuroscience to the effect that the mind does contain models of certain aspects of itself alongside models that capture how the rest of the world works.[54]

How, then, does the conscious self manage to evade the infinite regress that threatens self-referential systems? One possibility here is for the self's model of itself to eschew unlimited recursion by employing exclusively subpersonal (subconscious) processes, which are barred from calling upon the conscious virtual entity that emerges from their interaction. Another possibility might be to allow some recursion, assuming that this approach to predictive control is not only safer than full recursion but also more effective than no recursion at all. Alternatively, unlimited recursion may be allowed after all, if convergence can be guaranteed, in other words, if after a few iterations the planned action computed by the model remains unchanged to within a given tolerance.

Merely managing to avoid infinite regress does not, however, bring the self-modeling controller out of the woods yet. A real-life controller would have a goal: to promote desirable outcomes (a goal that Conant and Ashby left out of their formulation of good regulator, so as to make their theorem more widely applicable).[55] To that end, the controller comes up with a number of possible interventions and predicts the likely outcome for each of these, using its model of self-in-the-world. It then evaluates the predicted outcomes in terms of their desirability, so that it can rank them and choose the best action. But desirability depends on the structure and dynamics of the controller's self, which may well be changed by the action: what it holds to be desirable now may be spurned by its newer version, once the prospective intervention is completed. In an attempt to catch up with itself, the controller computes a new set of candidate actions and evaluates *their* consequences, only to face the same predicament all over again.

To break this vicious circle, the controller may pretend that its self is not affected by its actions. It would thus settle for one round of prediction and evaluation, cutting short the incipient infinite regress. Or, the controller may run as many exploratory iterations as necessary until an action plan is found that is projected to be acceptable to the version of its self that would emerge as a result of that action. A middle way would be to search for a solution for as long as possible, then accept and act on the best candidate plan from among those considered.

A conscious self, which is the virtual pilot constructed by a total flight simulator, shares the self-modeling controller's predicament — without necessarily being aware of any difficulties that arise out of it. Indeed, in certain tasks, such as homeostasis maintenance, there are no difficulties to begin with, because the controller's action needs to restore the system to a state that is invariant in its desirability. Thus, when I contemplate whether or not to get a drink of water to quench my thirst, I do not get distracted by the thought that I may not want a drink once I am no longer thirsty.

Allostasis, or goal-oriented regulation, is different: when trying to decide whether or not to taste a raw oyster for the first time in my life, I may find it hard to reliably project the hedonic state I'll be in after the oyster goes down my gullet. The difficulties that people have with this kind of prediction, called affective forecasting, are well-known.[56] The problem is particularly dire when the prediction concerns a remote future that depends in part on one's choice in the present situation. Think of trying to decide whether or not to embark on a costly multi-year project, such as getting a college degree. Will I still be happy in four or five years with my present decision to go for it? It has been pointed out that such long-term projects are generally difficult to manage because they are outside the scope of our evolutionary heritage.[57] That may be so, but there is also the objective difficulty that we have just identified: it inheres in the formal computational nature of the self-in-the-world setup, taken together with the Good Regulator theorem. To know oneself — the classic exhortation to wisdom[58] — is hard enough (or else we all would be wise); to know one's future self is harder.

Phenomenal Content and Phenomenal Affect

Getting to know the world that we are born into is an open-ended project — not the least because it keeps changing and we, who are part of it and are trying to figure out what it is and what we are, change right along with it. While certain necessary knowledge is handed down to us in the genome and the developmental sequences that it sets in motion and controls, most of what we need to know we must learn from experience. The computational algorithms for reinforcement learning make it possible to do so without explicit instruction, by noting the consequences of applying the newly acquired knowledge.

In learning to model the world, the gold standard for progress is especially easy to formulate. The prediction imperative requires that the model be used to forecast the sensory data. The predicted pattern can then be compared to the actual one, and the hypotheses about the hidden causes of the sensory input adjusted, so as to minimize the difference. Mathematically, this difference — the *prediction error* — is nothing but a measure of the model's free energy,[59] which, as noted earlier, caps the model's surprise at the sensory input. Because its minimization is the key to the model owner's well-being and, indeed, survival, the dynamics of prediction error feeds right into the agent's virtual clearing house for vitally important stuff: its consciousness.

All phenomenal conscious experience is CONTENT, tinged with AFFECT.

Being conscious consists of experiencing certain *content* (as in registering information and drawing distinctions) integrally with its accompanying *affect*, or feelings. The emerging consensus in *computational phenomenology* is that what we experience consciously as aspects of affect — valence, emotions, and mood — is in fact the dynamics of prediction error. The key characteristics of this dynamics are the sign (positive or negative) and magnitude of the prediction error and its rate of change, as well as its precision (inverse of estimated variance). The challenge of computational phenomenology is to clarify how these numbers relate to the lived experience. To fully meet this challenge, we must understand the "what," the "why," and the "how" of each variety of experience in question: the computations that constitute it, the reasons for which they are undertaken, and the processes that implement them.

These questions are typical of the study of systems that deal in information, for which the computational understanding must be sought on several levels. On the most basic implementation level, phenomenal experience consists in the dynamical system's state ascribing an intrinsically distinct trajectory — one that tells itself apart from other trajectories — through the space of possibilities. In creatures like us, this would be the space of possible brain states. Studying consciousness exclusively in terms of the brain processes that implement it would, however, make for a very narrow understanding (this is one reason for the brain having largely been left out of this book, the other being the distinct possibility of having consciousness without any biology). To broaden this understanding, we must also figure out what computations the system carries out by virtue of undergoing the dynamics it does, and why.

What might the answers to these questions look like? At this moment, my experience includes an appreciation of the coffee I am having (just brewed from freshly ground peaberry beans, cooperative-grown in Custepec valley, Chiapas, Mexico). Let's take a closer look first at the content of this experience. As a combination of smell and taste, it evokes other smells and tastes. Although I may not be enough of a coffee expert to explicitly discern the "notes of dark chocolate, pear, and brown sugar" listed on the package, my present experience would not be the same without prior exposure to chocolate, pears, and sugar. Indeed, my memories, such as they are, of everything that I ever ate, drank, or smelled, as well as of places where I have had those experiences, is what imparts meaning to my present experience.

This is the sense in which all our experiential content is *relational*. In computational terms, the phenomenal qualities that I am experiencing right now are accounted for by the similarities between, on the one hand, the shape of the activity-space trajectory presently unfolding in my brain, and, on the other hand, memory traces of activities evoked in the past by other coffees, chocolate, pears, apples, sugar, honey, visits to Mesoamerica, opinions on Chiapas-style cooperative agriculture and anarcho-syndicalism, etc. As I put it elsewhere, all representation is representation of similarities; and because all experience, being *about* something, is representational, it too consists in similarities — to other experiences.[60]

The relational nature of phenomenal content has been noted early on in the history of scientific psychology, by William James. In an essay titled *Does 'consciousness' exist?* published in 1904, James offers a description of consciousness that echoes the computational formulation just stated. Consciousness, James writes (using italics for emphasis), does exist — just not in the same sense that objects or substances do:

> *The peculiarity of our experiences, that they not only are, but are known, which their 'conscious' quality is invoked to explain, is better explained by their relations — these relations themselves being experiences — to one another.*

The idea, then, is that conscious experiences are relational. Their existence is more like that of dance moves than of the dancer. A dance move is defined conditionally, in relation to other moves, rather than on its own and in absolute terms — just as a conscious experience only makes sense in relation to other experiences.[61]

According to the doctrine of dependent origination or *pratītyasamutpāda*, a key tenet of Buddhist philosophy, all dharmas (phenomena) arise and exist conditionally — that is, solely in dependence upon other dharmas. This insight is crucial for understanding the concept of universal emptiness or *śūnyatā*, as declared in the Prajñāpāramitā Sutra: what everything is empty of is unconditional or absolute existence.

A more recent philosophical analysis by Daniel Dennett takes the relational nature of phenomenal experience for granted.[62] No matter that *qualia* — the "atoms" of phenomenal experience, such as the redness of a rose or the tartness of a lemon — appear to intuition as the epitome of all that is absolute and intrinsic. This intuition is misleading and is easily dispelled by comparing a quale to others like it: the redness of a rose to that of arterial or venous blood; the tartness of a lemon to that of a lime or a chokecherry. "Properties," Dennett writes, "that "seem intrinsic" at first often turn out on more careful analysis to be relational. [...] The public versions of perceptual "qualia" all *seem* intrinsic, in spite of their relationality."

Tasmāc Śāriputra, śūnyatāyāṁ na rūpaṁ, na vedanā, na saṁjñā, na saṁskārāḥ, na vijñānam
(Therefore, Śāriputra, in emptiness there is no form, no feeling, no perception, no volition, no consciousness)
— From the *Prajñāpāramitā-Hṛdayam Sūtra*

Western philosophy was by no means the first to have converged on the idea of relationality of all qualities. This idea is the conceptual foundation of the thesis of dependent origination or *pratītyasamutpāda* in philosophical Buddhism, according to which nothing exists unconditionally: as it is usually expounded, "this exists because of that; if that ceases to exist, so does this." One of the most prominent applications of the logical argument that underlies this thesis is to the intuitive concept of the self, which too is thereby revealed to be empty of unconditional or intrinsic existence.[63]

Notably, the dependent origination argument and its logical consequences are less radical than popularly perceived. The assertion of emptiness, or *śūnyatā*, means that what everything is empty of is only *unconditional* existence. By pointing out that the self can only be defined relationally, the emptiness doctrine presages the computational account, according to which the self is not a "thing" but a process — an essentially and fundamentally representational one. The self exists in relation to whatever it is that the representations that constitute it are representations of.

In an exciting twist of the plot, foundational theories that sound a lot like the thesis of dependent origination have recently been gaining prominence in physics, in the work of Lee Smolin, Carlo Rovelli, and others. Here is Smolin's summary of his approach, which is motivated by the key cosmological consideration that a theory of the universe cannot invoke any entity that is outside it:

> In a completely relational theory, a universe with no fixed background [that is, no "outside"] can be described as an inter-nested complex of dynamical subsystems, each one of which can be thought of as being described in terms of its interactions with the rest. We talk about this in terms of the *view* a subsystem has of its surroundings, by which we mean the information available to it of those other subsystems it interacts with most strongly.[64]

Rovelli makes a similar theoretical move in dealing with the notorious problem of observer in quantum mechanics, which has to do with the apparent conflict between the inherently probabilistic nature of quantum variables, as described by the "wave function," and the situations when these variables take on definite values:

> [S]omething must determine *when* a variable has a value. The textbook answer is 'when we measure it.' This obviously makes no sense, because the grammar of Nature certainly does not care whether you or I are 'measuring' anything. Measurement is an interaction like any other. Variables take value at any interaction. [...] The answer of relational QM [quantum mechanics] is that the variable a of the system S actualized in the interaction with S' takes value *with respect* to S, but not with respect to S'. This is the core idea underlying the 'relational' interpretation of QM.[65]

The double action of relational QM theory with regard to consciousness is worth noting. On the one hand, relational QM does away with the need for a conscious observer who would "collapse" the wave function; on the other hand, it helps us place consciousness squarely within the physical universe. In any case, if Nature is indeed relational in this sense, we should be only too happy to settle for a similarly relational theory of consciousness.

Coming back to my coffee experience, whose content is relational as discussed, we can now take up the computational questions — the "what," the "why," and the "how" — about the *affect* that it carries. Like any other affective experience, my enjoyment of coffee is characterized by valence (in this case, positive) and arousal (mild). The same kind of answer offered earlier to the implementation-level ("how") question applies here: valence and arousal too can be interpreted in terms of activity-space trajectories. Roughly, negative or positive valence corresponds, respectively, to aversive or attractive regions of activity space, while arousal has to do with the strength of aversion/attraction (the steepness of the gradient). But what computational purpose do these trajectory-space structures serve, and why?

The predictive processing framework offers convincing computational hypotheses addressing such higher-level questions of what and why. The explanations it proposes are based on the idea that the system's overarching goal in learning and in the control of behavior is to reduce the level of prediction error afforded by the generative model it employs — and with it, the level of surprise at the imminent developments. One such explanation, called the Affective Inference Theory, equates valence with the expected rate of prediction error reduction.[66]

On this account, my experience with Chiapas-grown coffee feels positive because its flavor and aroma fit my expectation of what coffee should be like, so that my brain's prediction error rapidly decreases (after a few such experiences, as my taste adapts, the rate of the decrease may be reduced, and with it the magnitude of the still positive valence). In contrast, the brew concocted from a stale grind of burnt beans, poured from a jug that's been sitting on the hot plate for hours, elicits a highly negative experience, because its vileness beats even my originally low expectations from an office mail-room or convenience store coffee.

I set down my mug and click on the Twitter tab in the browser window to catch up on the news. (As you can see, I do not have the willpower to forestall distractions by doing my writing on an offline computer.) Getting the news from a microblogging social medium is fun: it lets one guess what happened by drawing inferences from comments made by accounts I respect, instead of listening to corporate media sock puppets. The news this time is, unfortunately, not fun: the county health agency reports a spike in local Covid cases. While the gloominess of the news is not surprising (it fits what the actuarial statisticians call the secular trend), its specific content is. As such, it ratchets up the errors associated with my predictions of the probability of catching the virus in the classroom, in the swimming pool, or on the bus. The valence of this bit of news is definitely negative.

It is worth noting that the Affective Inference Theory accounts not just for the polarity and magnitude of valence associated with an experience, but also for its "aboutness" (what philosophers term *intentionality*): the property of being directed at a certain object or aspect of the state of affairs in the world. In the case of having coffee, it is the black liquid in the cup that my predictions are about, as their effectiveness can clearly only be evaluated by taking a sip. And there is no way I could mistake my coffee for the cause of my bad feeling about the news I just glimpsed.

As a little preview of the next chapter, note the role language plays here. The character of my experience of the news that I read differs from that of the coffee that I smell and taste: it links to my memories and conceptual representations not just viscerally, but also symbolically. When one is threatened to be shown fear in a handful of dust, it's what the dust signifies and portends that one dreads. This fearsome power of language — a toolkit for conceptual sophistication, for abstract reference, and for making people do your bidding — is available to all, even if not all of us have, like T.S. Eliot, honed our tools to a lethally sharp point.

What if I decide to avoid all surprise and make my world fully predictable by retreating into a safe, offline space in my home, never to see the light of a screen again? Evolution must be unkind to animal species that opts for this means of managing negativity, anxiety, and shock, yet aspires to a mobile existence: unless you're a barnacle, you would eventually choose uncertainty and danger over terminal boredom — and the likelihood of being eaten by something that you were unaware and thus unafraid of.

Creatures that stick to the exploitation of the resources available in their present and thus relatively familiar, unsurprising, and (so far) safe location, at the expense of venturing out to explore less familiar places, do not make good learners. Effective learning by definition requires exposure to the unknown, and thus a proper balance between exploitation and exploration.[67] To keep this balance, the agent must consider surprise simultaneously on a range of time scales. In the short term, the reduction of surprise is needed, for reasons explained above; in the longer run, the temporary increase in surprise during exploration makes it possible to learn a better model of the world — and, as we shall see in chapter 7 to build a better world.

Another useful perspective on the "dark room" problem is offered by self-evidencing, which underlies our bodily inner sense. Recall that maximizing the evidence for our model of the world requires minimizing surprise. In planning behavior, this implies that we are compelled to choose actions that minimize expected surprise. However, expected surprise just *is* uncertainty. This means that we are, by virtue of our existence, curious creatures, who consider novelty and the resolution of uncertainty to be inherently attractive — and therefore unsurprising.[68] It is because I am born to be surprised that finding myself in a dark room would be surprising, and therefore uncomfortable, to me. Speaking of affect, this suggests that the states of high uncertainty that (we anticipate) will resolve are exactly those in which we experience excited anticipation and pleasure. Conversely, in a state of uncertainty that (we anticipate) will not resolve, we experience agitated angst that, if it endures, generates stress, expressed physiologically as allostatic load.

A long-term success in driving down the prediction error (despite occasional fluctuations arising from unexpected "teachable moments") results in an unspecific affect that is experienced as positive *mood.* Along the way, action selection on part of the agent is biased both by its mood and by the more fleeting emotions, evoked by the transient dynamics of prediction error reduction. Or, things may not be going well because of bad luck, or because the agent is inept — or is set up for failure by a system that is too unpredictable or outright hostile to be learned and confronted effectively. In that case, the long-term average of the rate of prediction error reduction may be too low for the agent to feel good about itself, driving it to negative mood, and perhaps depression.

Virtual Reality, Real Suffering

Seeing emotional valence and mood for what they are — manifestations of the dynamics of prediction errors, generated by the agent's model of the world and of itself — helps us understand the dialectic of happiness and suffering, the two inseparable ingredients of the constant stream of appraisals churned out by the self's performance monitoring system. The computational need for this system to stay calibrated ensures that even during the best of times our phenomenal experience is suffused with both positive and negative feelings.

The complex shape of our affect is due to the multiplicity of dimensions and the range of time scales on which self-evaluation happens, as well as to the constantly shifting reference points for each dimension and time scale. An organism whose performance evaluation system comes equipped with a fixed setting for neutral affect is likely to be out-competed by one in which this setting tracks recent history. This is why there is always room for both positivity and negativity in our ongoing experience.

The dynamic adjustment of the reference point for subjective well-being is also why the popular view of happiness as an unalloyed good, to be pursued relentlessly, is deeply mistaken. We are physiologically incapable of staying happy once a goal has been achieved. There is a good evolutionary — and ultimately, computational — reason for that: we come from a long line of creatures who lived long enough to have and raise offspring *because* they would not rest on their laurels after a win.[69]

The evolutionary explanation of the abiding presence of unhappiness in our conscious lives does not, of course, make it easier to bear. If it were possible for a conscious agent to switch off emotions at will, they would lose their effectiveness. For a similar reason, there is little comfort in our hard-won understanding that all emotions are "mere" computations, insofar as they reside in, and are experienced by, the virtual self, the ghost pilot in the body's machine. As with all other indicators on the pilot's "system status" display, including the very reality one finds oneself in and must act on, emotions must be taken at face value, or else the pilot will sink into the morass of second-guessing its own senses. Our senses are *transparent* to us, in that what we see is not nerve signals that encode patterns of light, but rather the world. Our emotions, too, are transparent: what we feel is not nerve signals that encode prediction errors, but rather happiness — and misery. The reality we inhabit is virtual, but the suffering that is to be found therein is very real.

The Third Revolution

The arising of self-aware individuals is the third one in the cascade of consciousness revolutions that this book documents. To the non-egoic "effective self" of chapter 2 — a self-organizing, persistent local causal domain — self-awareness adds dimensions of body ownership, perspective, and agency. Two more ingredients are still missing: language and society, so closely intertwined in humans as to be inseparable. These will be explored in chapters 4 and 5.

Meanwhile, the result is something we can almost recognize ourselves in. A self-aware self that is speechless and asocial is what I am when on one of my solo hikes in the desert — for as long as I manage to disengage from inner speech and from thinking about family, friends, and students, and can simply *be* then and there. Being human means, however, that such disengagement can only be temporary; someone who can sustain it must, as Aristotle put it in *Politics*, be either a beast or a god.

In any case, even with the rest of the human universe successfully set aside, for good or even just for the moment, the recluse's freedom is illusory. The great web of cause and effect in which every self is firmly embedded governs all. The twin rewards of enlightenment afforded by self science are the awareness of the dependent origination of all things and a recognition of the roles of chance and necessity in every self's decisions.

Insofar as one's decisions are not without consequences, it is possible to learn from experience, which is indeed what creatures that are subject to evolutionary pressure do. Per the Good Regulator theorem, this is best done by modeling as many of the aspects of the environment as necessary and practical. The benefit of such modeling is the ability to predict where the world is headed. The computational framework that focuses on predictive processing has been shown effective in explaining many key particulars of phenomenal selfhood, from its immanent subjectivity and the slipperiness of the mind's *I* under introspective scrutiny, to the nature of affect as dynamics of prediction error.

While the affective dimensions of conscious experience reflect the varying tensions between predicted and actual representations of the world (including the self's internal goings-on), the contents of consciousness express the manifold relationships between the system's present dynamics and its memories of past experiences. Content-wise, the self is constituted by memories. The richer your life, the more complex the you that exists in the moment.

The inseparability of one's ongoing, momentary, phenomenal experience and one's memories, or life experience, is an important consolation of old age. Still, affect-wise, experiential complexity is a mixed blessing. Records of one's past experiences can be positive or negative; and so one's happiness is always shot through with sorrow at its incompleteness, imperfection, and fleetingness. Then, of course, there is sorrow in its own right. While some sorrows may be eased by hoping that they too shall pass, not all do.

That suffering — that is, consciously realized unhappiness of the inescapable variety — is a key to human phenomenology and the human condition is indicated by the preoccupation on part of all major world religions with soteriology, or doctrines of salvation. It is the third consciousness revolution that unleashed the full human potential for suffering, so it stands to reason that it will take another revolution to roll back the "progress" achieved by this one. The prospects for such a revolution, as well as other roads to freedom from suffering, will be the subject of Part II of this book (chapters 6 and 7).

That part is still a long way off, though, and we do need a dash of hope to sustain us till we get there. Hope, such as it is, will be found in the proper understanding of the causes of suffering. The first distinction to make is between avoidable and unavoidable varieties. While the latter is inherent in how mind-like computations work and how minds are shaped by evolution, the former is up to us — or rather, up to our fellow humans. By being of the right mind, and by taking the right actions, we should collectively be able to alleviate personal suffering, and perhaps do away entirely with its social and systemic causes. The tool that must be bent to this purpose — language — is what I turn to next.

❧ ❧ ❧

Notes

1. St. Anselm's argument dates back to the 11th century CE. For an overview of ontological arguments for the existence of God, see (Oppy, 2020). Williams (1962) offers a fascinating discussion of the concept of existence. In this connection, see also the Buddhist doctrine of emptiness — that is, the emptiness of absolute (as opposed to dependent) existence (Priest, 2009) — which comes up later in this chapter.

2. Margolis (1964) offers an analysis of the *cogito* argument, as does Unger (1979), who opens with a quote from A. J. Ayer:

> The sentence 'I exist', in this usage, may be allowed to express a statement which like other statements is capable of being either true or false. It differs, however, from most other statements in that if it is false it can not actually be made. Consequently, no one who uses these words intelligently and correctly can use them to make a statement which he knows to be false. If he succeeds in making the statement, it must be true.

3. The phrase *sentio, ergo sum* is due to Johann Gottfried Herder (Sugiyama, 2019). For a psychological angle on this notion (in the context of motivation), see (Cantril, 1967). Modern computational treatments of the sense of existence are based on the concept of embodiment (Craig, 2002; Hohwy and Michael, 2017; Seth and Tsakiris, 2018).

4. Here is how Locke puts it: "This Source of Ideas, every Man has wholly in himself ... And though it be not Sense, as having nothing to do with external Objects; yet it is very like it, and might properly enough be call'd internal Sense." (Locke, 1690, II.1.iv). For a brief review of the medical history of the concept of self-feeling, see (Rzesnitzek, 2014). The health angle is discussed by Koban et al. (2021). Northoff and Panksepp (2008) expand the discussion to a "trans-species concept of self." The identification by Solms (2019, p.7) of consciousness with "felt uncertainty" connects the concept of self-feeling with the predictive processing framework, which I discuss at length later in this chapter.

5. Regarding the delusion of self-nonexistence (Cotard's syndrome), see (Radovic, 2017).

6. An even better analogy would be to the "dead man's switch" that is sometimes installed on buses and trains and is designed to switch the engine off, unless the driver keeps pressing a special pedal.

7. A useful introduction to Friston's work, which focuses particularly on the intimate connection between the computational underpinnings of life and of minds, can be found in (Friston, 2013, 2010); see also Kirchhoff (2018).

8. Regarding self-evidencing, see (Friston, 2018; Hohwy, 2016).

9. This, of course, is the evolutionary take on existence; subjective views, which weigh the agreeable aspects of existence (such as they are) against the suffering that invariably accompanies it, may differ. See (Edelman, 2020), ch. 12 (*Existence*) and ch. 32 (*Suffering*).

10. There is no end to the philosophical debate about whether or not a thermostat can be conscious. I will not indulge in it here.

11. Regarding the computational commonalities between "life-like" and "mind-like" systems, see (Seth and Tsakiris, 2018, p.979): "The deep physiological roots of instrumental inference gesture towards a [...] connection between 'life and 'mind' [...] Our view suggests [...] that there are intimate connections between the functional imperatives imposed by our physiological reality, by the drive to stay alive that animates all living creatures, and the predictive machinery that implements instrumental interoceptive inference." Kauffman and Clayton (2006, pp.506-507) discuss the thermodynamics of minimally life-like "molecular autocatalytic systems." Solms and Friston (2018) argue that the minimal thermodynamic requirements for being alive, when considered together with the endogenous nature of consciousness, constitute "sufficient conditions for attributing feeling to being."

12. Kauffman and Clayton (2006, p.519) define autopoiesis in terms of "the concept of the cycle of work: the constrained release of energy, and the use of work to construct those very constraints." See also (Allen and Friston, 2018; Kirchhoff, 2018; Miguel Signorelli and Meling, 2021; Rudrauf et al., 2003).

13. The quote is from (Llinás and Roy, 2009, p.1305). In an earlier work, Llinás offers a similar formulation: "We know this centralization of prediction as the abstraction we call the 'self' " (Llinás, 2001, p.23).

14. Sensory substitution technology has been pioneered by Paul Bach y Rita; see (Bach-y-Rita, 1967; Bach-y-Rita and Kercel, 2003; White et al., 1970).

15. A sleeping brain too can create a virtual world, a different one, when it dreams. A classic philosophical parable here is Zhuangzi's butterfly dream (Cheng, 2014). See Ursula K. Le Guin's (1971) novel *The Lathe of Heaven* for a fascinating fictional treatment and (Edelman, 2011) for the computational considerations that apply to any attempt to distinguish between dreaming (or simulation) and reality.

16. The concept of efference copy originated with von Holst and Mittelstaedt (1950); see (Godfrey-Smith, 2019) for an illuminating discussion of the broader implications of reafference. Merker (2005) ties it to the biology of subjectivity, selfhood, and action.

17. For a clinical presentation of the room tilt illusion, see (Sierra-Hidalgo et al., 2012); a detailed case study can be found in (Arntzen and Alstadhaug, 2020). The vestibular origins of this family of illusions are discussed in (Malis and Guyot, 2003). Tekgün and Erdeniz (2021) report the relative weighting of vestibular and visual cues.

18. The source of the quote is *La dioptrique* by Descartes (1637). Note that the lens of the eye produces an image that is inverted both upside down and left to right.

19. Dennett (1988) gets it just right: "Only a very naive view of visual perception could sustain the idea that one's visual field has a property of right-side-upness or upside-downness *independent of one's dispositions to react to it* — "intrinsic right-sideupness" we could call it." Dolezal (1982) is a book-length report of experience with adaptation to inverting goggles.

20. The rubber hand illusion was first reported by Botvinick and Cohen (1998).

21. The same computational principles that make the rubber hand illusion possible apply also to tickling; Finlay (2019, p.4) describes "psychophysical demonstrations of how much and on what dimensions a pattern of stimulation can be distorted from one's own movement to be adequately unpredictable to no longer count as 'self' and become tickle-inducing."

22. Here is a partial list of the recent reports on how our bodily self-perception can be subverted by combined visual and tactile stimulation: (Blanke, 2012; Blanke and Metzinger, 2009; Blanke et al., 2002; Ehrsson, 2007; Guterstam et al., 2020; Lenggenhager et al., 2007). For an auditory counterpart, see (Lesur et al., 2021). Disorders of own body perception, such as somatoparaphrenia, are discussed in (Vallar and Ronchi, 2009).

23. The alien hand syndrome is discussed in depth by Biran and Chatterjee (2004). Seghezzi et al. (2021) review disorders in the sense of agency.

24. Here is minimal agency in the words of Kauffman and Clayton (2006, p.505): "We shall call a system able to act on its own behalf *an autonomous agent*." Similarly, Biehl et al. (2016) write: "First, define the spatiotemporal patterns which are suitable to represent both living (bacteria, animals, plants) and non-living (rocks, crystals) persistent objects. Then further classify those patterns into classes exhibiting features of agents such as perception, action and goal-directedness. Spatiotemporal patterns that satisfy all criteria will represent agents."

25. Buhrmann and Di Paolo (2017, p.218) argue in favor of the presence of agency and value judgment in single-cell animals:

> These cells are agents in the sense that they can regulate interactions with their environment in a way that support their continued existence (e.g. by regulating osmotic pressure). Otherwise neutral external affairs thus gain a valenced status with respect to the cell. What is good or bad for it is not arbitrarily defined by an external observer, but is intrinsically determined by its processes of self-constitution.

26. The comparator model of agency and body ownership is from (Gallagher, 2000, Fig.1). Buhrmann and Di Paolo (2017) offer a broader definition, according to which the sense of agency is "a phenomenological consequence of enacting sensorimotor schemes." For a robotics perspective, see (Hafner et al., 2020). The phenomenology of agency can be modulated by high-level cognitive processes: for instance, van der Weiden et al. (2011) found that causal knowledge affects self-agency experience.

27. I am using "random" here to stand in for anything that is unpredictable. The prime example here is the randomness found in quantum physics and stochastic thermodynamics, but at the scale of biotic self-organisation too there is unpredictability, which is due to chaos (e.g., the exponential divergence of trajectories) in dynamical systems. Interestingly, this source of randomness arises from the particular nonlinearities that define biological systems, namely, solenoidal or conservative dynamics of oscillations, biorhythms, and life-cycles. Technically, these break detailed balance and thereby underwrite non-equilibria of the sort found in life.

28. See (Dennett, 1984). I have offered a similar take on free will in (Edelman, 2008, section 10.1) and (Edelman, 2020, ch.14).

29. The quote is from (Lem, 1984, p.88).

30. See the story about Hodja Nasreddin's cemetery encounter in (Edelman, 2020, ch.14).

31. And perhaps not even that: a fully deterministic physics, which has no place for any randomness, has been outlined by Gerard 't Hooft (2019).

32. Here is Crutchfield (1994, p.26) on agency: "The goal for the agent is to detect the "hidden" states $\mathbf{S} = \{S_0, S_1, \ldots, S_{V-1}\}$ in its sensory data stream that can help it predict the environment. The states so detected will be called "causal" states. For discrete time series a causal state is defined to be the set of subsequences that renders the future *conditionally independent* of the past." Compare this to Rovelli's (2020) remarks on the thermodynamics of agency: "The agent is here simply the place where we arbitrarily decide to start the sequence of correlations described by the laws we are interested in: it is, in other words, where we ignore previous physical links." Note that the use of a personal pronoun ("we") in this formulation is not essential: it can be rephrased in entirely impersonal terms.

33. As Wittgenstein (1958, 621) noted, "... [V]oluntary movement is marked by the absence of surprise."

34. Thought insertion is a symptom in schizophrenia (Gallagher, 2015; Pickard, 2010) and psychotic states (López-Silva, 2018). For a broader philosophical perspective on the sense of agency in thinking, see (Doyle, 2020).

35. In contrast, Block (2011) claims that "perceptual consciousness overflows cognitive access."

36. Importantly, unlike in the Cartesian Theater scenario lampooned by Dennett (1991), where a homunculus is watching a presentation staged by the sensory processes, here all the components of what I call the "spectacle in the head" — including the audience — are virtual.

37. The feeling of the freedom of will has been called "the mind's best trick" by Wegner (2003), who also offers a theoretical account of the experience of conscious will (Wegner, 2004).

38. This quote is from Still et al. (2012, p.120604).

39. For a comparison of model-free and model-based reinforcement learning, see (Dayan and Berridge, 2014). A key reference for causal learning is (Pearl, 2018).

40. Some reviews of the power of generative models and the related Bayesian inference framework in cognition can be found in (Friston, 2009; Kersten et al., 2004; Tenenbaum et al., 2011).

41. I first came across a reference to the original Good Regulator paper (Conant and Ashby, 1970) while reading (Metzinger, 2003). Scholten (2010) has a detailed analysis of Conant and Ashby's original proof, as well as a discussion of the relationship of their concept of optimal regulation to homeostasis and to the reduction of uncertainty. Eykhoff (1994) examines various control schemes in light of the Good Regulator theorem. Raut et al. (2020) have recently invoked it in the context of brain dynamics.

42. A definition of the concept of entropy can be found in any textbook on thermodynamics or information theory. Sinai (1959) applies it to dynamical systems.

43. In the by now familiar terminology of conditional probability, we can say that the agent's states, which are internal to the blanket, are conditionally independent of the (rest of the) world's states, as long as the blanket states are known. A recent must-read on Markov blankets is (Bruineberg et al., 2021), where an important distinction is made between inference with a model and inference in a model ("The agent does not have a model of its environment that it uses to perform inference, but rather the agent is a model of its environment (Friston 2013; Bruineberg, Kiverstein, and Rietveld 2018; Friston 2019; Baltieri and Buckley 2019).") Note that the concept of Markov blanket is symmetrical with respect to the inside and the outside of the set of blanket

variables. Insofar as it does not bestow a privileged status to "subject" over "object" variables, it is thus somewhat reminiscent of the non-duality thesis in Yogācāra and related schools of Buddhism (e.g., Mackenzie, 2018).

44. Regarding surprise minimization via the free energy route, see (Buckley et al., 2017; Friston, 2010). Note the relationships to the concepts of homeostasis and autopoiesis (Friston, 2013).

45. Limanowski and Blankenburg (2013) highlight "the potential power of the FEP [Free Energy Principle] in explaining the mechanisms underlying minimal selfhood and its key constituents, multisensory integration, interoception, agency, perspective, and the experience of mineness."

46. Kirchhoff and van Es (2021) mount a "universal ethology challenge to the free energy principle"; they also discuss the concept of good regulator as it applies in this context.

47. As pointed out by Karl Friston (personal communication), what minimizing surprise through action looks like from the point of view of a behavioral psychologist is exactly reinforcement learning — a key concept, which appears in this book throughout chapters 1–3. In other words, if we associate surprising outcomes with those outcomes that a creature like me tends to avoid, by minimizing surprise it will look as if I am seeking reward or averting surprising (bad), aversive outcomes. On this view, reinforcement learning is an emergent property of self-evidencing when one replaces rewarding outcomes with the kind of outcomes that I find unsurprising and expect myself to encounter.

48. This evolutionary argument is presented in (Edelman, 2015).

49. "The subjectivity of subjective experience" is the title of (Metzinger, 2004); see also (Metzinger, 2018).

50. Metzinger (2003, p.553) writes: "I claim that phenomenal first-person experience works like a *total flight simulator*."

51. I offer this conclusion tentatively, as it is based on an informal appeal to the Good Regulator theorem. A formal proof would have to be based on an expanded version of the theorem, which would explicitly allow for the regulator's inclusion in the regulated system.

52. For a definition of the product of dynamical systems, see for instance (Nghiem, 1973).

53. The dizzying prospect of infinite regress that is inherent in any attempt to include a model of the self in one's model of the world may be an answer to what Chalmers (2018) called the "meta-problem of consciousness." This consists in accounting for the widespread intuition (which I do not share) that reducing consciousness to properties of matter presents a uniquely "hard problem" (Chalmers, 1995). Clark et al. (2019) offer an alternative answer to the meta-problem, based on the view of the brain as a hierarchical inference and prediction engine.

54. See (Deane, 2021; Hofstadter, 1979; Metzinger, 2003, 2018).

55. See (Scholten, 2010, pp.15-17).

56. Regarding affective forecasting, see, e.g., (Pilin, 2021; Wilson and Gilbert, 2005). Its role in major depression is discussed in (Kube et al., 2020).

57. The disconnect between the evolutionary roots of, and the present demands on, the mechanisms that support our sense of well-being is a central point made by Nesse (2004). For a discussion of happiness and regret, see (Edelman, 2020, ch.15 and ch.28).

58. Hence the inscription *gnothi seauton* ("know thyself") above the entrance to the Delphic oracle's cave. Similar sentiments are found in classical Chinese philosophy, e.g., (Lao Tze (Laozi), 1904, 33): "knowing others is wisdom; knowing the self is enlightenment"; and Buddhism, where mindfulness is defined as "complete self-mastery by means of self-knowledge" (Eliot, 1921, X).

59. Fernandez Velasco and Loev (2020, p.5) note that "a fundamental, assumption of PP [the predictive processing theory] is that free-energy corresponds to the amount of prediction error (Friston, 2009, p.293)."

60. The notion that "representation is representation of similarities" is from (Edelman, 1998). It is worth stressing that all phenomenal experience is essentially representational (Metzinger, 2004) (see also Kozuch, 2020).

61. The William James quote is from (James, 1904).

62. See (Dennett, 1988). A fundamentally relational computational approach to consciousness is stated and defended in (Tsuchiya and Saigo, 2021). A thorough neurophilosophical review of relational structuralism as a framework for consciousness research can be found in (Lyre, 2022).

63. Concerning the doctrine of emptiness in Buddhism, see (Priest, 2009). Scharfstein (1998, p.464) notes parallels between this idea and Hume's philosophy. For a brief overview of some related concepts, see (Edelman, 2020, ch.10).

64. Smolin (2020) continues: "We posit that space is not present at the fundamental level of description, but is emergent, together with the relative positions of the various subsystems." Note that here I am not concerned with ontology, merely with epistemology, which corresponds roughly to the Yogācāra stance on these matters (e.g., Mackenzie, 2018).

65. For an brief overview of relational quantum mechanics, see (Rovelli, 2018).

66. Fernandez Velasco and Loev (2020, p.23) write:

> AIT's [Affective Inference Theory] central tenet is that the valence of affective experiences corresponds to Expected Rate, so that positive (negative) feelings correspond to Expected Rate being positive (negative). According to AIT, Expected Rate is the mark of the affective.

In addition, the precision (inverse variance) of the error estimate is important: according to Fernandez Velasco and Loev (2020, p.6), it "serves then as a form of weighting when the system is deciding the degree to which the current hypothesis should be updated in light of incoming Error — highly precise Error will play a larger role in driving hypothesis updating. Accordingly, the brain is always trying to optimise precision, which is conducive to Error minimisation over time. In the PP literature, this process of precision-optimisation is what defines attention (Hohwy, 2012)."

67. Regarding the need to balance exploration and exploitation in behavior, see (Hills et al., 2015). An account of this balance in terms of predictive coding can be found in (Hesp et al., 2021, p.407).

68. The deep conceptual connections among uncertainty, surprise, and the exploration-exploitation dilemma are discussed in (Barto et al., 2013; Schwartenbeck et al., 2013). In a recent study that combined imaging with computational modeling, Cockburn et al. (2022) had participants choose between options that varied in terms of novelty, uncertainty, and expected reward, while the task horizon was being manipulated between trials. The results suggested three independent computational processes at work: reward seeking, novelty seeking, and uncertainty aversion. These were subject to timing constraints: on the one hand, the subjects adaptively evaluated the benefit of reducing uncertainty and were less likely to choose uncertain options as the prospective advantage of additional information declines; on the other hand, despite the inherent uncertainty of novel options, these were pursued regardless of the task horizon (Cockburn et al., 2022, p.2701).

69. This is the central thesis of *The Happiness of Pursuit* (Edelman, 2012).

References

Allen, M., & Friston, K. J. (2018). From cognitivism to autopoiesis: Towards a computational framework for the embodied mind. *Synthese, 195*, 2459–2482.

Arntzen, K., & Alstadhaug, K. B. (2020). Room tilt illusion and subclavian steal — a case report. *BMC Neurology, 20*, 369.

Bach-y-Rita, P. (1967). Sensory plasticity: Applications to a vision substitution system. *Acta Neurologica Scandinavica, 43*, 417–426.

Bach-y-Rita, P., & Kercel, S. W. (2003). Sensory substitution and the human-machine interface. *Trends in Cognitive Sciences, 7*, 541–546.

Barto, A., Mirolli, M., & Baldassarre, G. (2013). Novelty or surprise? *Frontiers in Psychology, 4*, 907.

Biehl, M., Ikegami, T., & Polani, D. (2016). Towards information based spatiotemporal patterns as a foundation for agent representation in dynamical systems. In *ALIFE 2016: The Fifteenth International Conference on the Synthesis and Simulation of Living Systems* (pp. 722–729). Cambridge, MA: MIT Press.

Biran, I., & Chatterjee, A. (2004). Alien hand syndrome. *Archives of Neurology, 61*(2), 292–294.

Blanke, O. (2012). Multisensory brain mechanisms of bodily self-consciousness. *Nature Reviews Neuroscience, 13*, 556–571.

Blanke, O., & Metzinger, T. (2009). Full-body illusions and minimal phenomenal selfhood. *Trends in Cognitive Sciences, 13*, 7–13.

Blanke, O., Ortigue, S., Landis, T., & Seeck, M. (2002). Stimulating illusory own-body perceptions. *Nature, 419*, 269.

Block, N. (2011). Perceptual consciousness overflows cognitive access. *Trends in Cognitive Sciences, 15*(12), 567–575.

Botvinick, M., & Cohen, J. (1998). Rubber hands 'feel' touch that eyes see. *Nature, 391*, 756.

Bruineberg, J., Dołęga, K., Dewhurst, J., & Baltieri, M. (2021). The emperor's new Markov blankets. *Behavioral and Brain Sciences, 22*(45), e183. https://doi.org/10.1017/S0140525X21002351

Buckley, C. L., Kim, C. S., McGregor, S., & Seth, A. K. (2017). The free energy principle for action and perception: A mathematical review. *Journal of Mathematical Psychology, 81*, 55–79.

Buhrmann, T., & Di Paolo, E. (2017). The sense of agency — a phenomenological consequence of enacting sensorimotor schemes. *Phenomenology and Cognitive Science, 16*, 207–236.

Cantril, H. (1967). Sentio, ergo sum: "motivation" reconsidered. *The Journal of Psychology, 65*, 91–107.

Chalmers, D. (2018). The meta-problem of consciousness. *Journal of Consciousness Studies, 25*(9–10), 6–61.

Chalmers, D. J. (1995). Facing up to the problem of consciousness. *Journal of Consciousness Studies, 2*, 200–219.

Cheng, K.-Y. (2014). Self and the dream of the butterfly in the *Zhuangzi*. *Philosophy East and West, 64*(3), 563–597.

Clark, A., Friston, K., & Wilkinson, S. (2019). Bayesing qualia: Consciousness as inference, not raw datum. *Journal of Consciousness Studies, 26*, 19–33.

Cockburn, J., Man, V., Cunningham, W. A., & O'Doherty, J. P. (2022). Novelty and uncertainty regulate the balance between exploration and exploitation through distinct mechanisms in the human brain. *Neuron, 110*, 2691–2702.

Conant, R. C., & Ashby, W. R. (1970). Every good regulator of a system must be a model of that system. *International Journal of Systems Science, 1*, 89–97.

Craig, A. D. (2002). How do you feel? Interoception: The sense of the physiological condition of the body. *Nature Reviews Neuroscience, 3*, 655–666.

Crutchfield, J. P. (1994). The calculi of emergence: Computation, dynamics, and induction. *Physica D, 75*, 11–54.

Dayan, P., & Berridge, K. C. (2014). Model-based and model-free Pavlovian reward learning: Revaluation, revision, and revelation. *Cognitive, Affective, & Behavioral Neuroscience, 14*, 473–492.

Deane, G. (2021). Consciousness in active inference: Deep self-models, other minds, and the challenge of psychedelic-induced ego-dissolution. *Neuroscience of Consciousness, 7*(2), niab024.

Dennett, D. C. (1984). *Elbow room: The varieties of free will worth wanting*. Cambridge, MA: MIT Press.

Dennett, D. C. (1988). Quining qualia. In A. J. Marcel & E. Bisiach (Eds.), *Consciousness in modern science* (pp. 42–77). Oxford: Clarendon Press.

Dennett, D. C. (1991). *Consciousness explained*. Boston, MA: Little, Brown & Company.

Descartes, R. (1637). *La dioptrique*. Leyden: Ian Maire.

Dolezal, H. (1982). *Living in a world transformed: Perceptual and performatory adaptation to visual distortion*. New York: Academic Press.

Doyle, C. (2020). The sense of agency and the epistemology of thinking. *Erkenntnis, 87*, 2589–2608.

Edelman, S. (1998). Representation is representation of similarity. *Behavioral and Brain Sciences, 21*, 449–498.

Edelman, S. (2008). *Computing the mind: How the mind really works*. New York, NY: Oxford University Press.

Edelman, S. (2011). Regarding reality: Some consequences of two incapacities. *Frontiers in Theoretical and Philosophical Psychology, 2*, 44.

Edelman, S. (2012). *The happiness of pursuit*. New York, NY: Basic Books.

Edelman, S. (2015). Varieties of perceptual truth and their possible evolutionary roots. *Psychonomic Bulletin and Review, 22*, 1519–1522.

Edelman, S. (2020). *Life, death, and other inconvenient truths*. Cambridge, MA: MIT Press.

Ehrsson, H. H. (2007). The experimental induction of out-of-body experiences. *Science, 317*, 1048.

Eliot, C. (1921). *Hinduism and Buddhism: An historical sketch.* London: Routledge & Kegan Paul.

Eykhoff, P. (1994). Every good regulator of a system must be a model of that system. *Modeling, Identification and Control, 15*, 135–139.

Fernandez Velasco, P., & Loev, S. (2020). Affective experience in the predictive mind: A review and new integrative account. *Synthese, 198*, 10847–10882.

Finlay, B. F. (2019). The neuroscience of vision and pain: Evolution of two disciplines. *Philosophical Transactions of the Royal Society B, 374*, 20190292.

Friston, K. (2013). Life as we know it. *Journal of The Royal Society Interface, 10*(86), 1–12.

Friston, K. (2018). Am I self-conscious? (Or does self-organization entail self-consciousness?). *Frontiers in Psychology, 9*, 579.

Friston, K. J. (2009). The free-energy principle: A rough guide to the brain? *Trends in Cognitive Sciences, 13*, 293–301.

Friston, K. J. (2010). The free-energy principle: A unified brain theory? *Nature Neuroscience, 11*, 127–138.

Gallagher, S. (2000). Philosophical conceptions of the self: Implications for cognitive science. *Trends in Cognitive Sciences, 4*, 14–21.

Gallagher, S. (2015). Relations between agency and ownership in the case of schizophrenic thought insertion and delusions of control. *Review of Philosophy and Psychology, 6*, 865–879.

Godfrey-Smith, P. (2019). Evolving across the explanatory gap. *Philosophical Theory and Practical Biology, 11*, 1.

Guterstam, A., Larsson, D. E. O., Szczotka, J., & Ehrsson, H. H. (2020). Duplication of the bodily self: A perceptual illusion of dual full-body ownership and dual self-location. *Royal Society Open Science, 7*, 201911.

Hafner, V. V., Loviken, P., Pico Villalpando, A., & Schillaci, G. (2020). Prerequisites for an artificial self. *Frontiers in Neurorobotics, 14*, 5.

Hesp, C., Smith, R., Parr, T., Allen, M., Friston, K. J., & Ramstead, M. J. D. (2021). Deeply felt affect: The emergence of valence in deep active inference. *Neural Computation, 33*, 398–446.

Hills, T. T., Todd, P. M., Lazer, D., Redish, A. D., & Couzin, I. D. (2015). Exploration versus exploitation in space, mind, and society. *Trends in Cognitive Sciences, 19*, 46–54.

Hofstadter, D. R. (1979). *Gödel, Escher, Bach: An eternal golden braid.* New York: Basic Books.

Hohwy, J. (2012). Attention and conscious perception in the hypothesis testing brain. *Frontiers in Psychology, 3*, 96.

Hohwy, J. (2016). The self-evidencing brain. *Noûs, 50*(2), 259–285.

Hohwy, J., & Michael, J. (2017). Why should any body have a self? In F. De Vignemont & A. J. T. Alsmith (Eds.), *The subject's matter* (pp. 363–391). Cambridge, MA: The MIT Press.

James, W. (1904). Does 'consciousness' exist? *Journal of Philosophy, Psychology, and Scientific Methods, 1*, 477–491.

Kauffman, S. A., & Clayton, P. (2006). On emergence, agency, and organization. *Biology and Philosophy, 21*, 501–521.

Kersten, D., Mamassian, P., & Yuille, A. (2004). Object perception as Bayesian inference. *Annual Review of Psychology, 55*, 271–304.

Kirchhoff, M. D. (2018). Autopoiesis, free energy, and the life-mind continuity thesis. *Synthese, 195*, 2519–2540.

Kirchhoff, M. D., & van Es, T. (2021). A universal ethology challenge to the free energy principle: Species of inference and good regulators. *Biology & Philosophy, 36*, 8.

Koban, L., Gianaros, P. J., Kober, H., & Wager, T. D. (2021). The self in context: Brain systems linking mental and physical health. *Nature Reviews Neuroscience, 22*, 309–322.

Kozuch, B. (2020). No pain, no gain (in Darwinian fitness): A representational account of affective experience. *Erkenntnis, 85*, 693–714.

Kube, T., Schwarting, R., Rozenkrantz, L., Glombiewski, J. A., & Rief, W. (2020). Distorted cognitive processes in major depression: A predictive processing perspective. *Biological Psychiatry, 87*, 388–398.

Lao T. (Laozi). (1904). *The book of the simple way*. London: Philip Wellby. Translated by W. G. Old.

Le Guin, U. K. (1971). *The lathe of heaven*. New York, NY: Avon Books.

Lem, S. (1984). Chance and order. *The New Yorker, 59*, 88–98.

Lenggenhager, B., Tadi, T., Metzinger, T., & Blanke, O. (2007). Video ergo sum: Manipulating bodily self-consciousness. *Science, 317*, 1096–1099.

Lesur, M. R., Bolt, E., Saetta, G., & Lenggenhager, B. (2021). The monologue of the double: Allocentric reduplication of the own voice alters bodily self-perception. *Consciousness and Cognition, 95*, 103223.

Limanowski, J., & Blankenburg, F. (2013). Minimal self-models and the free energy principle. *Frontiers in Human Neuroscience, 7*, 547.

Llinás, R., & Roy, S. (2009). The 'prediction imperative' as the basis for self-awareness. *Philosophical Transactions of the Royal Society London B, 364*, 1301–1307.

Llinás, R. R. (2001). *I of the vortex*. Cambridge, MA: MIT Press.

Locke, J. (1690). *An essay concerning human understanding*. https://www.gutenberg.org/files/10615/10615-h/10615-h.htm

López-Silva, P. (2018). Mapping the psychotic mind: A review on the subjective structure of thought insertion. *Psychiatric Quarterly, 89*, 957–968.

Lyre, H. (2022). Neurophenomenal structuralism. A philosophical agenda for a structuralist neuroscience of consciousness. *Neuroscience of Consciousness, 8*(1), 1–18.

Mackenzie, M. (2018). The Yogācāra theory of three natures: Internalist and non-dualist interpretation. *Comparative Philosophy, 9*(1), 18–31.

Malis, D.-D., & Guyot, J.-P. (2003). Room tilt illusion as a manifestation of peripheral vestibular disorders. *Annals of Otology, Rhinology, and Laryngology, 112*, 600–605.

Margolis, J. (1964). I exist. *Mind, 73*(292), 571–574. New Series.

Merker, B. (2005). The liabilities of mobility: A selection pressure for the transition to consciousness in animal evolution. *Consciousness and Cognition, 14*, 89–114.

Metzinger, T. (2003). *Being no one: The self-model theory of subjectivity*. Cambridge, MA: MIT Press.

Metzinger, T. (2004). The subjectivity of subjective experience: A representationalist analysis of the first-person perspective. *Networks, 3–4*, 33–64.

Metzinger, T. (2018). Splendor and misery of self-models: Conceptual and empirical issues regarding consciousness and self-consciousness. *ALIUS Bulletin, 1*(2), 53–73. Interviewed by J. Limanowski and R. Millière.

Miguel Signorelli, C., & Meling, D. (2021). Towards new concepts for a biological neuroscience of consciousness. *Cognitive Neurodynamics, 15*, 783–804.

Moyal, R., Fekete, T., & Edelman, S. (2020). Dynamical emergence theory (DET): A computational account of phenomenal consciousness. *Minds and Machines, 30*, 1–21.

Nesse, R. M. (2004). Natural selection and the elusiveness of happiness. *Philosophical Transactions of the Royal Society B, 359*, 1333–1348.

Nghiem, D.-N. (1973). On the classification of dynamical systems. *Annales de l'Institut Henry Poincaré, Section B, 9*(4), 397–425.

Northoff, G., & Panksepp, J. (2008). The trans-species concept of self and the subcortical-cortical midline system. *Trends in Cognitive Sciences, 12*, 259–264.

Oppy, G. (2020). Ontological arguments. In E. N. Zalta (Ed.), *The Stanford encyclopedia of philosophy*. Metaphysics Research Lab, Stanford University.

Pearl, J. (2018). Theoretical impediments to machine learning with seven sparks from the causal revolution. arXiv:1801.04016.

Pelevin, V. (1998). Tai Shou Chuan USSR. In *A werewolf problem in central Russia and other stories* (pp. 80–94). Cambridge: New Directions Publishing Corporation. Translated by A. Bromfield.

Pickard, H. (2010). Schizophrenia and the epistemology of self-knowledge. *EuJAP, 6*(1), 55–74.

Pilin, M. A. (2021). The past of predicting the future: A review of the multidisciplinary history of affective forecasting. *History of the Human Sciences, 34*(3–4), 290–306.

Priest, G. (2009). The structure of emptiness. *Philosophy East & West, 59*, 467–480.

Radovic, F. (2017). The sense of death and non-existence in nihilistic delusions. *Phenomenology and Cognive Science, 16*, 679–699.

Raut, R. V., Snyder, A. Z., & Raichle, M. E. (2020). Hierarchical dynamics as a macroscopic organizing principle of the human brain. *Proceedings of the National Academy of Science, 117*(34), 20890–20897.

Rovelli, C. (2018). Space is blue and birds fly through it. *Philosophical Transactions of the Royal Society A, 376*, 2017.0312.

Rovelli, C. (2020). Agency in physics. https://arxiv.org/abs/2007.05300.

Rudrauf, D., Lutz, A., Cosmelli, D., Lachaux, J.-F., & le van Quyen, M. (2003). From autopoiesis to neurophenomenology: Francisco Varela's exploration of the biophysics of being. *Biological Research, 36*, 27–65.

Rzesnitzek, L. (2014). Narrative or self-feeling? A historical note on the biological foundation of the "depressive situation". *Frontiers in Psychology, 5*, 9.

Scharfstein, B. (1998). *A comparative history of world philosophy: From the upanishads to kant.* Albany, NY: SUNY Press.

Scholten, D. L. (2010). A primer for Conant & Ashby's "Good-Regulator Theorem" (Unpublished ms). http://www.goodregulatorproject.org/images/A_Primer_For_Conant_And_Ashby_s_Good-Regulator_Theorem.pdf

Schwartenbeck, P., Fitzgerald, T., Dolan, R. J., & Friston, K. (2013). Exploration, novelty, surprise, and free energy minimization. *Frontiers in Psychology, 4*, 710.

Seghezzi, S., Convertino, L., & Zapparoli, L. (2021). Sense of agency disturbances in movement disorders: A comprehensive review. *Consciousness and Cognition, 96*, 103228.

Seth, A. K., & Tsakiris, M. (2018). Being a beast machine: The somatic basis of selfhood. *Trends in Cognitive Sciences, 22*(11), 969–981.

Sierra-Hidalgo, F., de Pablo-Fernández, E., Herrero-San Martín, A., Correas-Callero, E., Herreros-Rodríguez, J., Romero-Munõz, J. P., & Martín-Gil, L. (2012). Clinical and imaging features of the room tilt illusion. *Journal of Neurology, 259*, 2555–2564.

Sinai, Y. G. (1959). On the notion of entropy of a dynamical system. *Doklady of Russian Academy of Sciences, 124*, 768–771.

Smolin, L. (2020). On the place of qualia in a relational universe. https://philarchive.org/archive/SMOTPO-3

Solms, M. (2019). The hard problem of consciousness and the free energy principle. *Frontiers in Psychology, 9*, 2714.

Solms, M., & Friston, K. (2018). How and why consciousness arises: Some considerations from physics and physiology. *Journal of Consciousness Studies, 25*(5–6), 202–238.

Still, S., Sivak, D. A., Bell, A. J., & Crooks, G. E. (2012). Thermodynamics of prediction. *Physical Review Letters, 109*, 120604.

Sugiyama, T. (2019). Herder on "sentio, ergo sum": Seen from his remarks on the color harpsichord. In Y. Suzuki, K. Nakagawa, T. Sugiyama, F. Akiba, E. Maestri, I. Choi, & S. Tsuchiya (Eds.), *Computational aesthetics* (chap. 3, pp. 43–51). Tokyo: Springer.

't Hooft, G. (2019). Free will in the theory of everything. In F. Scardigli, G. 't Hooft, E. Severino, & P. Coda (Eds.), *Determinism and free will: New insights from physics, philosophy, and theology* (pp. 21–48). Cham: Springer.

Tekgün, E., & Erdeniz, B. (2021). Influence of vestibular signals on bodily self-consciousness: Different sensory weighting strategies based on visual dependency. *Consciousness and Cognition, 91*, 103108.

Tenenbaum, J. B., Kemp, C., Griffiths, T. L., & Goodman, N. D. (2011). How to grow a mind: Statistics, structure, and abstraction. *Science, 331*, 1279–1285.

Tsuchiya, N., & Saigo, H. (2021). A relational approach to consciousness: Categories of level and contents of consciousness. *Neuroscience of Consciousness, 2021*(2), niab034.

Unger, P. (1979). I do not exist. In G. F. Macdonald (Ed.), *Perception and identity* (chap. 10, pp. 235–251). London: Macmillan.

Vallar, G., & Ronchi, R. (2009). Somatoparaphrenia: A body delusion. A review of the neuropsychological literature. *Experimental Brain Research, 192*, 533–551.

van der Weiden, A., Aarts, H., & Ruys, K. I. (2011). Prime and probability: Causal knowledge affects inferential and predictive effects on self-agency experiences. *Consciousness and Cognition, 20*, 1865–1871.

von Holst, E., & Mittelstaedt, H. (1950). The reafference principle interaction between the central nervous system and the periphery. In *Behavioral physiology of animals and man: The collected papers of Erich von Holst* (vol. 1). Coral Gables, FL: University of Miami Press. First published in *Die Naturwissenschften* 37:464-476. Translated by Robert Martin (1973).

Wegner, D. M. (2003). The mind's best trick: How we experience conscious will. *Trends in Cognitive Sciences, 7*, 65–69.

Wegner, D. M. (2004). Precis of *The Illusion of Conscious Will*. *Behavioral and Brain Sciences, 27*, 649–692.

White, B. W., Saunders, F. A., Scadden, L., Bach-y-Rita, P., & Collins, C. C. (1970). Seeing with the skin. *Perception and Psychophysics, 7*, 23–27.

Williams, D. C. (1962). Dispensing with existence. *The Journal of Philosophy, 59*(23), 748–763.

Wilson, T. D., & Gilbert, D. T. (2005). Affective forecasting: Knowing what to want. *Current Directions in Psychological Science, 14*, 131–134.

Wittgenstein, L. (1958). *Philosophical investigations* (3rd edn.). Englewood Cliffs, NJ: Prentice Hall. Translated by G. E. M. Anscombe.

4 Speech and Sign

The dawning of magic. A shared toolbox. Writing. Stories and culture. Narrative selves. Power, persuasion, and propaganda. Manifesto destiny. The fourth revolution.

> Thus, we have in language the fact (and this seems obvious to me) that words began, in a sense, as magic.
>
> *This Craft of Verse: The Charles Eliot Norton Lectures 1967-1968*
> — JORGE LUIS BORGES

> To obtain forgiveness for its sins mankind needs only to declare them for what they are.
>
> — KARL MARX[1]

The Dawning of Magic

COULD the long emergence of consciousness described so far in this book have all taken place in the absence of as much as a squeak of language? We are a voluble species, and speech and sign are so central to the experience of being human that many people still believe that only humans are really conscious because only humans can express themselves (overtly or to themselves in their heads) in words. And yet, none of the levels of consciousness that I have discussed so far — not even self-consciousness, let alone minimal selfhood or selfless awareness — requires language.[2]

In contrast, language does need consciousness.[3] To be sure, all the low-level mechanics of its production (such as the choice of words and of utterance structure) and reception (understanding the utterance that is heard or seen) are taken care of by fully automatic, subconscious processes, as is the learning of language in children. It is the ultimate *use* that language is put to that necessitates conscious focus.

To a proficient user, language as transparent as water is to fish. Like the rest of high-level cognition, it is mediated by structured and potentially very complex representations, which, however, remain behind the scenes most of the time. On the production side, content that is properly structured for conveying the intended meaning (more about which in a moment) comes to hand as needed, typically without the speaker having to invest any conscious effort. It is the shaping of the meaning that may require conscious concentration. On the reception side, the structured content remains transparent, so that the meaning imputed by the listener (which may or may not correspond closely to the

S. Edelman, *The Consciousness Revolutions*,
https://doi.org/10.1007/978-3-031-24012-6_4

speaker's intention) is experienced directly, except when the regular interpretation process miscarries and consciousness is called upon to save the day (as it is in all behavioral emergencies).

This is why the expression "stream of consciousness" is something of a misnomer: free association is what language sounds like precisely when there is no intent to communicate and no coherent and conscious selection of content is being made. The utterances produced under those conditions may be locally well-structured, but they do not stand in any clear correspondence to the utterer's objectives, because none are brought to bear on the production. Language cannot be intentional when the conscious self, the virtual pilot that runs the show, is sidelined (or absent, as in dissociated states[4]).

Intentional action is, of course, what selves have evolved to support in the first place. Language is, however, special in that it affords a uniquely effective means for the self's intention to infiltrate another pilot's production, so as to make things happen there that would not have happened otherwise — such things, for instance, as an increase in the likelihood that the listener's pilot will direct its rig to make the speaker a sandwich.

Mauss (1902-3) was surely right in his view that magical ideas are a category of collective thought. ... Taking the notion of *mana* as his point of reference, he located the concept of spiritual force as the essence of magic, comparable to our concept of 'mechanical force', and was thus able to assimilate magic to the more general theme of causality. Magic was described as a 'gigantic variation on the principle of causality'.
— Stanley J. Tambiah (1968, p.203)

In a world devoid of magic, issuing commands to inanimate objects will not get you anywhere. In particular, it is not possible to cause a sandwich to come into existence by talking to its ingredients. This is why for a long time after life first appeared on earth, if someone wanted a sandwich, they had to make it themselves. It was the evolution of language that made it possible for someone not just to signal their hunger, but also to specify the elaborate manner in which they wish it to be satisfied — and perhaps to *cause* someone to satisfy it.[5]

Actionable language works as intended by the speaker when it is convincing, or backed by power. For someone to make you a sandwich, they must be capable of doing it and amenable to persuasion and you must be perceived as deserving. As my second-century namesake, the Talmudic sage Rabbi Shimon bar Yochai, put it, it is the righteous ("those who do God's will") whose chores are carried out by others.[6] In a later section, I shall discuss less benign, and arguably more common, scenarios in which persuasion is replaced by an exercise of power.

Given how close language does come to what we call magic (very), and considering the number of species on this planet that have it (one), the question of its evolutionary origins is both intriguing and difficult. It is hard to apply science to singular cases, and evolutionary science is particularly challenging when faced with a complex trait that has few, if any, close parallels across species and environments. In the case of language, there would also appear to be too little data to bring to bear: language, spoken or signed, leaves no fossils. The dearth of data having proved insufficient to discourage idle speculation on such an exciting topic, the French Academy of Sciences declared in 1866, famously, that it would no longer consider papers on the origins of language.

The general lack of attention to language evolution persisted through most of the following century, in part because the theoretical framework that dominated linguistics from the 1950s onward, Chomsky-style formalism,[7] has been consistently averse to the idea that language is a product of the same evolutionary mechanisms as everything else in biology. As formalist linguistics started to give way to functionalist approaches in the 1990s (interestingly, about the same time when scientific study of consciousness finally became reputable[8]), work on language evolution intensified. Eventually, it would lead to a wide acceptance of the view that language arose and evolved because of its special and unique — near-magical, one might say — contributions to human culture and sociality.

It was the shared material culture in particular that many scholars now believe to have been driving the evolution of language. This theory has not been entirely unanticipated. Just as the French Academy was coming out against speculation on the origins of language, Marxists were indulging in it. They were doing it without the least evidential basis, in part because back in 1876 there was none to be had. On the theoretical front, psychology was still trying to differentiate itself from philosophy and linguistics as such had not been invented yet.

First labour, after it and then with it speech — these were the two most essential stimuli under the influence of which the brain of the ape gradually changed into that of man.
— Friedrich Engels (1949), first published in 1876

At present, computational psychology and linguistics offer a clear picture of the commonalities between the representational structures and processes underlying language and the rest of cognition. Likewise, neuroscience has identified the shared brain basis of language and other sequential behaviors, such as navigation and complex manual tasks. Finally, language is now widely thought of as being learned during development (rather than a ready-made grammar being built into the newborn brain and merely waiting to be "activated").

This last piece of the puzzle suggests that what must have evolved in our ancestors is the *capacity for learning language*, and that the complexity and the diversity of modern languages are both due to cultural evolution, which unfolds in parallel with the slower biological adaptation. According to one hypothesis, it was manual labor in a social setting — specifically, the teaching of stone tool-making — that created the conditions for language to enter the repertoire of human behaviors and to stay competitive over sufficiently long evolutionary time for anatomical adaptation to take place. Having first come together through opportune coupling of neural mechanisms of communication and of serial behaviour during tool-making skill-sharing and pedagogy, the capacity for language became entrenched in the brain, building up an innate substrate for learning language.[9]

Since it first emerged (arguably, as early as the Lower Palaeolitic, about a million years ago[10]), language has been co-evolving with its users, taking on a variety of functions that have long expanded beyond the teaching of toolmaking. Along the way, language itself has become the most powerful tool at our disposal. Used in its *conative* capacity, this tool allows us to try and impel others (and ourselves) to do our bidding, as in the sandwich-making scenario.

The conative function would not, however, be of much use, were it not possible to express what it is that one wants to be done. Unlike causing someone to move in a particular direction, which can be achieved with a sharp stick and well-aimed prodding, sandwich creation involves intricately structured concepts that absolutely require expressive language. This is the sense in which a *bàhn mí* sandwich is little like an Acheulean hand ax — only a little, because decent sandwich-making can be learned in a few hours, whereas it takes years of practice to master the art of Neolithic stone knapping.

The special power of language thus lies not just in its conative function, and not even in the capacity for reference and concept-building that it affords, but rather in the open-ended nature of that capacity. This open-endedness is fully revealed in poetics, which Borges called magic, and which linguist Roman Jakobson, in his paper *Linguistics and Poetics*, placed at the center of a diagram depicting the functions of language.[11] The special power is that of *metaphor*. All reference in language (except iconic, as in using braying to de-

Jakobson (1960) lists six functions of language: referential, emotive, poetic, phatic, conative, and metalingual.

note "mule") is non-literal, but it is the pervasive use and continued invention of metaphors that allows us to actualize its full potential. Call a really good *bàhn mí* a poem written in pork, and who knows where you end up: a journey of a thousand *li* begins with but a single bite.

A Shared Toolbox

On those journeys through the space of meanings, which is infinite because it is always under construction, you are never alone: all the functions of language are inherently social. We all start by learning language through interaction with caregivers and peers. All linguistic experience involves more than one individual at a time — or else one individual across multiple times, as when inner speech is used for thinking in the privacy of one's mind. And even then we reuse language created by, or in collaboration with, others. In these ways, the exclusively human language-shaped superstructure of one's conscious self binds it to other selves much more strongly than its lower strata, discussed so far in this book, can.

Because the primary function of consciousness is effective control, social interactions that are of any importance are likely to be actively regulated by the socially engaged individual's virtual pilot. Language boosts the pilot's power of controlling its own host's behavior, by enabling explicit reasoning and planning, and also the behavior of others, by planting in their heads ideas that affect *their* reasoning. What is new and different about doing all this with language, as opposed to the old, prelinguistic way, is that the ideas are now *structured* and thus literally exponentially more expressive.

Behavioral coordination among animals controlled by what I called in earlier chapters 'effective selves' is not evolutionarily novel, nor is it specific to humans. Think of a bison, which may slip into a daydream while grazing, but will most certainly become alert and aware of its surroundings if its herd comes under an attack of wolves. The herd's collective behavior in this situation is shaped by coordination among individual bison. The wolves, which hunt as a pack, do more than coordinate: they *cooperate* with pack mates, with tangible benefits to all.[12]

The tools afforded by language make cooperation in humans (which will be the theme of chapter 5) more sophisticated, and its implications farther-reaching, than in any other social species such as bison or wolves, or even eusocial ones such as termites or bees.[13] Although promoting cooperation has arguably been the primary driver of language evolution in the first place, the language toolbox proved to be just as useful for thinking as it is for working with others. At this moment, for instance, my overarching goal is to get my understanding of what language is across to you, the reader. However, as this understanding is as yet partial and imperfect, I must resort to the use of the very tool I am trying to grasp, apply it to itself, then couch the resulting insights, such as they are, in terms that would make sense to both of us, before committing it all to the page.

The ability of language to support *reference* (whether literal or metaphoric) is the real basis of its power. Pictorial or iconic reference, as in the case of a petroglyph of a bison on the sandstone wall of a Utah canyon, works by resemblance — assuming that its creator had not meant it to refer to something else. As to the word "bison," which does not resemble a bison in any way, all bets are off: the utterer's intention is integral to fixing whatever it is (if anything) that reference is being made to in the given context.

This suggests that the *meaning* that gets traded in language use cannot be reduced to a correspondence between the symbol (a sound or a non-acoustic gesture) and its "referent." Likewise, communication cannot be reduced to an exchange of prepackaged and encoded meanings, which merely need to be "decoded" on the receiving side. Meaning is not sent and received — it is evoked.

Human beings relate to each other not simply externally, like two billiard balls, but by the relations of the two worlds of experience that come into play when two people meet.
— R. L. Laing (1967, p.53)

Making complex things happen through the magic of language takes properly structured spells. Having a dedicated single-word spell for each combination of settings of the task parameters would require too many spells — as many as the product of the number of relevant settings of each parameter. Worse, it would make it impossible to improvise a spell for a novel combination of parameter settings, let alone accommodate the emergence of an entirely new parameter. A much more efficient way to grow the space of potential meanings is to derive the requisite complexity of expression from the *composition* of symbols.

All human languages rely on syntactically structured utterances. We generate these "on the fly" by composing morphemes, which are the building blocks of words, into larger units, or *constructions*, subject to conventional or usage-based probabilistic constraints. It is the structure of these larger units — recurring sequences of words, as well as more abstract patterns formed by categories of words and, recursively, of patterns — that serves as the source of constraints on composition, also known as "grammar."[14]

The meaning of an utterance — ultimately, its effect on the listener[15] — depends in complex ways on the words and other constructions that can be discerned in it. Once these are taken up by the listener's brain, they interact, again in potentially complex ways, with the ongoing activity there, which in turn depends on the listener's personal history and present mindset. The interlocutors can understand what is being said — and, no less importantly, why — only by taking into account, in addition to the stream of symbols that are being exchanged, non-symbolic characteristics such as tone, prosody, and register, as well as the broader context of the interaction and the participants' shared cultural and perhaps personal history.[16]

Writing

I, as I write this, am only a certainty that seeks out the words that are most apt to compel your attention. The proposition and a few muscular sensations, and the sight of the limpid branches that the trees place outside my window, constitute my current I.
— Jorge Luis Borges (1999b)

In certain subcultures, such as the academe, and in particular academic linguistics, the widespread reliance on writing has given people the wrong impression that language can be equated with text, and that using language is a largely solitary affair, such that the interlocutors' respective contributions to a conversation can be effectively analyzed and made sense of in isolation.[17]

It is easy to see where these misguided assumptions may have originated: written language is uncanny. If someone's speech can cast a spell over me, how much more magical is the same effect when it is achieved by means of cuneiform depressions on a clay tablet, splotches of ink on paper, or pixel patterns on a screen? The use of durable media instead of ephemeral sounds or hand signs exposes the reader to the machinations of a magician whose power looms all the larger because its wielder is absent from the scene.

Still, this power, like that of all language, rests neither in the physical text nor in its author alone. As already noted, language has evolved as a social phenomenon, is acquired through social interaction, and remains fundamentally a collective undertaking in all its daily uses. The quintessentially dialogic nature of language and its relationship with consciousness is forcefully brought out in this quote from M. Bakhtin's monumental work on Dostoevsky's poetics, whose central thesis it expresses:

> An idea lives not in the *isolated* consciousness of a person: if left exclusively within it, it degenerates and dies. An idea begins to live — that is, to take shape, develop, find and renew the manner of its expression in words, beget new ideas — only by entering into concrete dialogic relationships with others' ideas. A human thought becomes genuine only under the conditions of live contact with others' thoughts, embodied in others' voices, that is, in the consciousness of others as it is expressed in words. It is at the point of contact between voices-consciousnesses that an idea is takes on life and stays alive.[18]

Speaking of ideas staying alive: a text can work magic even when its author, the magician, is not merely absent, but dead. Bakhtin, who was born fourteen years after the death of Dostoevsky, had himself been long dead when the edition from which I translated the above passage came out. The cognitive mechanisms whereby language gives a voice to the dead are at work not just when we read a disembodied text: our memories of encounters with people, living or dead, engage the very same processes. When I think of my late father, I sometimes can almost hear his voice in my head.[19]

The remarkable thing about writing is that it allows people to fall under the spell of magicians whom they admire but have never met in person. For me, the long list of such wizards is led by the names of Ursula Le Guin and Jorge Luis Borges.[20] Of course, like all magic, or technology, a book can be put to an evil use, or it may simply have unpredictable effects on the reader. It was while reading an essay by Borges, *On the Cult of Books*, that I came across a quote from one of the early Christian church fathers who likened publishing a book to placing a sword in the hands of a child.[21] At the same time, written language, which is critically important in modern education, may be our only hope for eventually mobilizing massive enough support for positive change. This is why the best educators, such as Paulo Freire, dedicated their lives to teaching literacy — and writing:

> If we think of education as an act of knowing, then reading has to do with knowing. The act of reading cannot be explained as merely reading words since every act of reading words implies a previous reading of the world and a subsequent rereading of the world. There is a permanent movement back and forth between "reading" reality and reading words — the spoken word too is our reading of the world. We can go further, however, and say that reading the word is not only preceded by reading the world, but also by a certain form of writing it or rewriting it. In other words, of transforming it by means of conscious practical action. For me, this dynamic movement is central to literacy.[22]

Stories and Culture

It is because so much of the human world is made of stories that all cultures rely on storytelling in educating children. It doesn't stop there: while culture, like language, is something one starts to acquire from the earliest infancy, its continued learning never really ceases. The product of this process is an edifice that overlays the physical world, which we perceive through the lens of culture, and often replaces it, as when the natural landscape gives way to cultivated fields, pasture, villages, and cities, all owing their existence to language. The world that most of us inhabit is thus literally shaped by the stories we tell: every agricultural practice, every architectural pattern, every engineering artifice, every piece of art, all entertainment, all gossip, all news, all politics — it's all shot through and held together with language.[23]

The world is made of stories, which bind us all together, and impossible stories are the best of all, for they bind us in impossible ways.
— Seth Dickinson (2018, ch.16)

Language shapes consciousness, among other ways, by allowing culture to grow cumulatively. Here, I am interested not at all in the material culture as expressed in technology and artifacts, and not so much in the "dead" culture that is stored in libraries and the data cloud, as in the active memories of conversations, of lessons learned and taught, and of stories heard and told. These may, of course, interact with, augment, and be enriched by the passively stored information, but unlike the latter, living language is part of our conscious experience, individual and collective. In particular, giving a book a proper reading (as envisaged by Freire, whom I quoted earlier) makes it come alive and become part of the living memory of the reader — and of the listeners, if read aloud. The accumulation and propagation of culture thus enrich the individual experience of its bearers (just like sensory experience is boosted by the person's hoard of memories), while building up their shared noosphere.

If we cared only for material culture or technological know-how, we would not be spending so much time and energy on telling and acting out stories. Intangible culture — a web of computational processes that run in the union of the virtual spaces spun by our selves — builds and maintains societies, which can also be radically changed, or even destroyed, by stories that resonate with people's individual consciousness and transform it. Consciousness-shaping epic language traditions, oral or written, are universally present in all human societies. The next chapter will take a closer look at the relationships between individual and *collective consciousness*.[24] Meanwhile, we can count some of the ways in which stories can bind people together by cultivating a society's collective consciousness, or drive them apart by stimulating clannishness, discord, and individualism.

The first of these is *ethnic consciousness*, promoted by epics such as the *Iliad* and the *Odyssey*, which Eric Havelock called "a body of invisible writing imprinted upon the brain of the community" in Homeric Greece;[25] the *Shahnameh* in ancient Persia;[26] and the *Mahabharata* in India.[27] In modern times, this type of consciousness developed into the ethos of nation-building and nationalism.

To be an Indian or simply to live in India at any period in her recorded history, is to open oneself to the benign moral influence of the two epics — the *Ramayana* and *Mahabharata*.
— Meera Prasannan (2019, p.1)

Whereas in the ancient epics gods and demons mixed freely with human heroes, the emergence of monotheism signaled a gradual shift to people-centered narratives. In these, the deity would still rule the world behind the scenes, but would typically not share the stage with the human actors. An example of such a narrative is the Hebrew Bible, which also marked a transition from ethnic to *national consciousness*, with the Book of Exodus in particular having provided the Israelite tribes with a foundation myth that combined sacred and secular elements. Because Judaism never encouraged proselytizing, it remained a predominantly ethnic religion. It anchored and sustained the Jews' conception of themselves for many centuries, until the *haskalah* or "education" movement led to the development of a parallel, secular Jewish consciousness.[28]

Certain other religions went on to transcend ethnic and nation-state boundaries to play a role in establishing the kind of *transnational consciousness* that characterizes the followers of the various denominations of Buddhism, Christianity, and Islam. The linguistic origins and subsequent history of these movements differ widely. Buddhism began as an oral tradition and remained largely so for hundreds of years, before the multitudes of stories that make up its canons were transcribed and translated into the languages of the people who professed it. The original languages of Christianity's sacred texts were the Hebrew of the Old Testament and the Koine Greek of the New; with time, translations have been made available: into Greek and Latin and Old Church Slavonic and, eventually, the many national languages of the Christian states. In contrast, Islam has always been firmly rooted in the original Arabic of the Koran, which all Muslims, regardless of their mother tongue or nationality, are required to use in prayer recitation and are encouraged to learn.

People do not, however, have to be religious to develop a collective consciousness that cuts right across national borders. It was a very recent political invention that caused borders, first in Europe and then all over the world, to be reified, reinforced with concrete and barbed wire, and rendered impassable except by permission of one's destination state (and, often enough, one's state of origin).[29] Enabling borderless peaceful coexistence of multiple ethnicities in adjacent or overlapping territories was the explicit intent behind the oral constitution that served as the foundational charter of the Haudenosaunee Confederation — an outstanding example of epic language promoting *transnational secular consciousness*.[30] This brings to mind another document explicitly designed to encourage people to transcend and indeed abolish all ethnic and national boundaries, one that derives a human-universal secular consciousness from the political-economic concept of social class. It is the *Communist Manifesto*, published in 1848 by Karl Marx and Friedrich Engels.

Narrative Selves

To appreciate the consciousness-shaping power of documents such as the Haudenosaunee constitution or the *Communist Manifesto*, we need to examine the role of stories in making up individual selves. Narratives are central to a person's self-concept and identity. The tangle of stories that we weave about ourselves structure our memories and help us make sense of our lives. The *narrative self* that grows out of these stories is a language-dependent, and therefore human-specific, addition to the already discussed levels and manifestations of consciousness, those that we share with other sentient beings.[31]

While a bat, like even the lowly lobster, has a biological self, it has no selfy self to speak of — no Center of Narrative Gravity, or at most a negligible one.
— Daniel Dennett (1991, p.448)

A person's narrative self has been figuratively described by Daniel Dennett (in a book-length story tantalizingly titled "Consciousness Explained") as their "center of narrative gravity" — the intersection, or common core, of all the stories in which they appear as a protagonist. And because a person (dead or alive) can appear in other people's stories, even if they don't share a language, the individual clouds of stories overlap and interpenetrate, so as to envelop all of humanity.

There is a certain thematic structure to our collective Cloud of stories. Part of it comes from the commonalities in different people's lived experiences: the daily routine, the kinds of events that occasionally break it, the life "milestones." Although the particulars in the stories we tell can be infinitely diverse, there are only so many meaningfully differing plots,[32] all of which have by this time made it into our collective cultural memory. It is from this common cultural stock of stories that we draw the building blocks of our individual narrative selves.[33]

The emergence of the narrative self as the core component of the human consciousness has been furthered by the shift, mentioned earlier, which took place over historical time in the common culture: from gods- and heroes-centered narratives to more down-to-earth, personal ones. In ancient Greece, this shift corresponded to an evolution of the relationship between epic culture and individual consciousness, in which the latter gradually took form and asserted its phenomenal independence. In his *Preface to Plato*, Eric Havelock described this evolution as a transition from what he called the "Homeric state of mind" — a "life without self-examination" — to "accepting the premise that there is a 'me', a 'self', a 'soul', a consciousness which is self-governing and which discovers the reason for action in itself rather than in imitation of the poetic experience."[34]

Explicit, language-mediated self-consciousness supplements the moment-to-moment control, provided by the pre-linguistic phenomenal awareness and effective self, with a sense of *personhood*, which encapsulates historical continuity and the evolution of one's experience over the lifetime.[35] Personal identity is not necessarily pondered on a daily basis, but can be pushed to the forefront of one's self-consciousness by certain social circumstances.[36] Multilinguals, for instance, may be reminded of the distinctness of their language-specific cultural personas, as control passes from one to another according to circumstances.[37]

In those who are minoritized by the narratives of the surrounding dominant culture, the sense of personal identity can be especially acute. As discussed, famously, by W. E. B. Du Bois, Frantz Fanon, and others, it can also then lead to a tension between one's self-perception and the prevailing objective conditions — a kind of *false consciousness.*[38] Unless held back by an overt mark of "otherness," such as the wrong skin color or incorrigible foreign accent, one may be tempted to reduce such tension, and the resulting existential angst and suffering, via cultural assimilation. An attempt to do so has been undertaken, on a massive scale, by the European Jews. That experiment ended in the gas chambers and crematoriums of Treblinka and Auschwitz.

As someone who was born in the USSR into a family of secular, thoroughly Russianized Jews (more about which in the next chapter), I participated in another well-known attempt at assimilation, only to be handed, time and again, proof that my people were still not welcome in our homeland of a thousand years. Some of us were prompted by this experience to move to a place where Jews would be the majority: the newly retaken "historic homeland" in Palestine. Most of those who did so settled comfortably into what they saw as the majority's prerogative: minoritizing and oppressing others.[39]

For the oppressed, the choice between, on the one hand, emigration — even if it does not entail oppressing others — and, on the other hand, assimi-

But almost always, during the initial stage of the struggle, the oppressed, instead of striving for liberation, tend themselves to become oppressors, or 'suboppressors'. The very structure of their thought has been conditioned by the contradictions of the concrete, existential situation by which they were shaped. Their ideal is to be men; but for them, to be a 'man' is to be an oppressor. This is their model of humanity.
— Paulo Freire (1970, p.11)

lation — even if, against all odds, it succeeds — is an evil one. For one thing, assimilation necessarily diminishes the shared treasure that is our species' collective cultural diversity. Perhaps more importantly, emigration and assimilation both amount to a surrender in the face of oppression. This is a matter not so much of honor as of a dangerous precedent being set: as the Russian proverb goes, give them a finger and they'll bite off your entire arm.

Millions of Diaspora Jews at the end of the 19th century faced the evil choice between attempting assimilation and joining the Zionist project: becoming a "regular people," masters of their own ethnostate. As all epic undertakings, Zionism was driven by a consciousness-shaping narrative — the very same one found in the Pentateuch and the Books of Prophets, which Theodor Herzl and his followers were now reading with an eye not to the past, but to the future: the building of *Altneuland*, the Old New Land.

The apparent success of the Jewish state-building undertaking and the tragic fate of those left behind in Europe have both overshadowed the existence of another narrative, which for a while carried the banner of a third way. In Herzl's time, it was represented by the ideas of Bernard Lazare, who rejected both the territorial-escapist and the assimilationist "solutions" to the Jewish question. Instead, Lazare envisaged an alliance between the Jewish poor and the other oppressed masses in Europe.[40]

This vision was also the driving force behind the General Jewish Labor Bund — a universalist socialist party ("Bund" means "union"), which opposed assimilation (and emigration), promoted secular ethnic self-consciousness, and aimed at joining forces with other workers' movements. The Bund, founded in 1897 in Lithuania, was active throughout the Russian Empire.[41] Following the October 1917 revolution, which it supported, it was swiftly suppressed by the Communist establishment. In Poland, the Bund remained active right up until the German invasion. Had the Bundists not perished — at the hands of the Bolsheviks during the terror and the purges of the 1920s; at the hands of the Nazis in the 1940s — their humanist vision would have helped bring about a better world for all, as imagined by another Jewish dreamer, Wolf Gordin:[42]

> And everyone will then shrug: how could it be that a small clique of oppressors lorded over the human masses, of which no less than half were women, of which the majority were oppressed nationalities, of which again the majority were youth and children, and of which, lastly, not only the majority, but pretty much everyone except the gang of bandits-ministers-commissars and billionaires was being personally oppressed by the state.

Power, Persuasion, and Propaganda

– Make me a sandwich!
– Get lost.
– sudo make me a sandwich.
– Coming right up.
— A Unix joke

In Unix, the programming environment that underlies much of the infrastructure of our personal and cloud computing, a user can summon the so-called command line interface to tell the system to do things like delete a file ("rm *filename*") or create a folder ("mkdir *foldername*"). Certain actions, such as messing with other users' files, are only available to the so-called superuser or root account, the most privileged one in the system. There is, however, a special command, sudo (short for "super-user do"), which allows any user who knows the root password to invoke the superuser privileges on a case-by-case basis, as in "sudo rm *someonelsesfile*".

By getting hold of your system's root password and amassing arcane knowledge about the available commands, backdoors, and exploits, you become a Unix wizard. In another self-contained (albeit fictional) domain, J. R. R. Tolkien's Middle Earth, wizards are likewise characterized by their extensive knowledge of spells — as are monsters. As the Company is fleeing from their foes through the Mines of Moria, Gandalf's door-shutting spell, intended to block the orcs' passage, is met with the advancing Balrog's "terrible" counter-spell, at which point Gandalf has no choice but "speak a word of Command" — presumably something along the lines of "sudo shut-the-door."[43]

In real life, the categories of wizards and monsters are defined by social power differentials and by the manner in which users, and super-users, apply their power. In certain situations — say, an encounter between an authoritarian parent and their child, or a boss and a wage worker, or a state and its subject, or a police officer and a normal person — a party having a power edge over you enables them to simply issue a command and expect it to be obeyed. In such cases, language plays the relatively simple role of a mere conduit of information: a specification of what it is that the recipient of the command must do (or else be smacked, or fired, or incarcerated, or shot).

Sociologists have long known, however, that such overt exercise of power represents only one of its many faces.[44] The second face of power is revealed in its agenda-setting prerogative, which is typically applied toward self-preservation. Thus, the corporate oligarchy that controls a "democratic" nation state may allow only those parties that report to the oligarchs to appear on the ballot. Or, the big employers may compel the politicians in their pay to pass laws preventing workers from organizing for better employment conditions. Typically, such laws are cast in a language that masks their content (let alone intent) from the casual consumer of the news, as when union-busting legislation is referred to as defending the "right to work."

The third face of power is even more far-reaching: it takes the form of manipulating people's consciousness, with the intent of making them more conforming and compliant. When this manipulation succeeds, the working-class voters will be unaware of the lack of any alternatives on the ballot to the political-economic status quo that keeps them unemployed and ready to work for any pay, while being denied health and family benefits. Here too language is all-important. In the country where I am writing these words, the United States of America, few people give any thought to the inhumaneness of treating housing or health "insurance" as commodities or privileges rather than basic rights. Nor, indeed, does the contemporary public discourse in this country, and most of the rest of the world, acknowledge the crooked conception of the human condition that lurks behind endlessly repeated and seemingly innocuous expressions such as "cost of living" or "living wage."

Mastery of language affords remarkable power.
— Frantz Fanon (1967, p.18)

It is their control over the language of public discourse that gives the power players in politics the ability to shape our collective consciousness. It does not matter if the political arena in question is a self-styled "democracy": in the best tradition of the concept of Newspeak from George Orwell's *1984*, the U.S. media happily entertain lethally euphemistic language such as "officer-involved shooting" for the execution of a citizen by the police, or "qualified immunity" for the shielding of police officers from even the kind of accountability that one may expect to be delivered by corrupt prosecutors, biased judges, and brainwashed jurors.[45]

The process of absolving the officers from accountability for their actions begins with the language in which the media cast police brutality. Here it is decried by June Jordan, in her "Poem about Police Violence" (1978):[46]

> I lose consciousness of ugly bestial rapid
> and repetitive affront as when they tell me
> 18 cops in order to subdue one man
> 18 strangled him to death in the ensuing scuffle
> (don't you idolize the diction of the powerful: *subdue*
> and *scuffle* my oh my) and that the murder
> that the killing of Arthur Miller on a Brooklyn
> street was just a "justifiable accident" again
> (Again)

Such violence at home — including the literal homes of the U.S. police officers, a disproportionate number of whom are domestic abusers[47] — often mirrors violence that is perpetrated by their country's military abroad. Similarities abound between the behavior of the police towards the people they are sworn to "serve and protect" and the behavior of the military towards those whom they are in the process of "liberating" from whoever the ruling class declares to be America's external enemies, no matter how far away they are geographically or how insignificant their military capabilities may be, compared to those of the country whose expenditure on "defense" outranks that of the next dozen countries combined.[48] This is how the phrase "surgical strike" becomes an accepted code for the bombing of a wedding party half way around the globe, which may or may not have been attended by a person whose namesake may have been placed on the relevant kill list by the democracy's military.

The technical term for the wholesale acceptance by the public of such language, and with it of the policies that it describes — or rather fails to describe — is *manufactured consent*.[49] The book by Edward Herman and Noam Chomsky, whose title first named this phenomenon, focused naturally on the U.S. as the country whose citizens are the least conscious of the realities behind its political system and the loyalties of the media that defend and extol it. In other countries, the means of keeping the public in compliance with the established order may differ, but the bottom line is the same. This explains several patterns in our behavior as a species: why we wage wars under imaginary or deplorable pretexts; why we do nothing about global warming despite the scientific consensus on its deadly momentum; and why we tolerate poverty and hunger, despite having enough resources and producing enough food to shelter and feed everyone.[50]

What one is or is not allowed to say or write in a given country is usually very telling with regard to the variety of political consciousness that prevails there. The freedom of expression that the rulers of the U.S. and other Western "democracies" grant to their citizens (witness my ability to publish this book) serves to show that they feel quite secure from the threat of subversion that they face from the likes of Chomsky, or from the few of us who think like him. In comparison, in the most populous pretend democracy on the planet, the "People's Republic" of China, the situation must be much more potentially volatile, or else its oligarchs would not be investing so much energy in suppressing every bit of speech, no matter how trivial, that may run counter to the official policies of the ruling Party, or threaten its absolute political and cultural hegemony.

The scope of the linguistic repression in China ranges from crackdowns on individual words that are perceived as subversive by the censors and banned from social media to country-wide, long-term language policy, which aims at promoting the dominance of Standard Chinese (*putonghua*) in public and private (family) usage over all ethnic languages, dialects, and topolects.[51] Given the central role of a person's language(s) in shaping their identity and in giving expression to their self-consciousness, this strategy clearly aims at policing not just what people dare to say, but also how and what they think of themselves, of others, and of the society they live in.

The invention of social media made it infinitely easier for politically like-minded people to connect with each other and organize. However, the internet, perhaps more than any other technology, is amenable to being coopted by individuals acting in bad faith, by corporations pursuing their agenda of amassing money and political power, and by the state, which exists to serve the interests of those with money and power. Given the unique technological means it offers for gathering and weaponizing personal data, the internet has become more than just another medium for manufacturing consent. It is used to drive the trivialization and fragmentation of civil discourse.[52] It is used to foster a paradoxical combination of know-nothing narrow-mindedness and parochialism on the one hand and on the other hand a global reach of conspiracy theories and straight-up fascism. And it helps amplify the dynamic that has all but destroyed the accepted notions of truth, both lay and scientific, ushering in the sad state of affairs that came to be called post-truth.[53]

The individual cannot bargain with the State. The State recognizes no coinage but power: and it issues the coins itself.
— Ursula K. Le Guin (1974, p.225)

Manifesto Destiny

We are not proposing any particular course of action. What we want is for people who read what we write to feel the scales falling from their eyes.
— Theodor Adorno and Max Horkheimer (2010, p.46)

In a letter that the 25-year old Karl Marx sent in 1843 to Arnold Ruge, with whom he co-edited the *Deutsch-Französische Jahrbücher*, he proposed that "mankind needs only declare" its sins to be forgiven.[54] But how can the people be made to see what is wrong with their collective predicament, when their upbringing and education, such as it is, condition them to conform, and when any dissent, if it is tolerated, is drowned in a constant onslaught of bad-faith language-twisting, relentless propaganda, truth-bending, and outright lies?

The format of the *Communist Manifesto*, which Marx published (together with Engels) five years later, is an answer of sorts to the all-important methodological question of how best to deploy language, not to suppress the people's consciousness or divert it, but to focus it on what it is that mankind must declare, as the first step on the road to making it better. In his introduction to a 2004 reprinting of the *Manifesto*, Len Findlay describes it as follows:[55]

> An interventionist genre designed to induce action in its readers (and in the less literate, to whom it was read aloud), its overt didacticism engages audiences who feel beleaguered. Accordingly, if — as some Victorians were fond of saying — the poor will always be with us, then the *Manifesto* will always be read. [...] The *Communist Manifesto*... continues to provide for the disadvantaged on the basis of a sociology of the stigma and economy of abuse. The stigma in question is poverty and the abuse the one that keeps elites in power.

In his letter to Ruge, Marx had declared his mission to be "the *ruthless criticism of the existing order*, ruthless in that it will shrink neither from its own discoveries nor from conflict with the powers that be."[56] That order — the system that perpetuates the stigma of poverty and the abuse of power by the elites — was, and still is, *capitalism.*

In the century and a half since its publication, the social-economic justice movements inspired by the *Manifesto* have caused the power of capitalism to be diminished, but they failed to destroy it. The great socialist revolutions, in Russia (1917) and China (1949), broke the system, but did not remove its foundation. Capitalism is made with, and preserved by, the power of the state, and while that remains it will endure. Indeed, in the long run both great revolutions (and countless lesser ones) grew into totalitarian caricatures of the regimes they had replaced. And so, the old *Manifesto* is as relevant as ever, and the capitalist state ever falls short of its goal of once and for all stamping out subversion: communist agitation against capitalism, and *anarchist* repudiation of the state itself and all the varieties of oppression that it engenders.[57]

Sauron was diminished, but not destroyed. His Ring was lost but not unmade. The Dark Tower was broken, but its foundations were not removed; for they were made with the power of the Ring, and while it remains they will endure.
— J. R. R. Tolkien (1954, *The Fellowship of the Ring*, Book 2, Ch. 2)

As far as language is concerned, this stalemate — not a draw, as the game is not over yet — is a testament to the limits of the power of persuasion and propaganda. But the stalemate cannot be broken by brute force: even if wholesale extermination of the "enemies of the state" is attempted, at least a few of the exterminators are liable to be contaminated with the seeds of dissent. Language remains, therefore, our species' only hope for a better future.

Realizing this hope is as difficult a task as any that we have ever faced collectively. The difficulty is hinted at by the question mark at the end of the title of the 1956 collection of dialogues between Theodor Adorno and Max Horkheimer on the prospects for political change: "Towards a new manifesto?"[58] Since mid-twentieth century, the task became, if anything, even more difficult. Whereas the original *Manifesto* stood out as a voice in the wilderness, the present proliferation of position papers, proclamations, pronouncements, policy statements, platforms, and programs suffices to bury all but the most incendiary *and* eloquent calls to action, even without the help of the capitalist state propaganda. Paraphrasing "ontological anarchist" Hakim Bey's memorable remark, what we should aim at is, perhaps, to *say* something children will remember all their lives.

... be ruthless, take risks, vandalize only what *must* be defaced, do something children will remember all their lives...
— Hakim Bey (1985, *Poetic Terrorism*)

The Fourth Revolution

It was a failure to *say* what needed to be said that cost Moses his entry ticket to the Promised Land. The Book of Numbers recounts how Moses, divinely instructed in the Zin wilderness to bring forth water out of rock by speaking to it, struck it instead. This did cause water to flow in abundance, yet the place came to be known as *Mei Meriva* (the Waters of Strife), for the contention sown among the people by Moses's choice. The prophet's job, after all, is all about having faith and using words, and Moses, this once, came up short on both counts.

And the LORD spake unto Moses and Aaron, Because ye believed me not, to sanctify me in the eyes of the children of Israel, therefore ye shall not bring this congregation into the land which I have given them. This is the water of Meribah; because the children of Israel strove with the LORD...
— Numbers 20:12,13

One has to be really good with language *and* have some serious leadership skills to be able to save lives, or change lives, with mere words. In his analysis of the story of Meribah, Rabbi Nathaniel Helfgot has argued that the real reason for the punishment of Moses was not the sin of disobeying a direct command from on high:[59]

> It was not the action per se, but rather what it represented or indicated as a test case of Moses' continued leadership of the new generation. Moses, Moshe Rabbenu, the master of all prophets and the leader of all leaders, was halted on the doorstep of the promised land and a new world, a place he could gaze at from afar but could not enter — *"Ki mineged tireh et ha'arets veshama lo tavo"* (Deut. 32:52).

Like so many other pieces of folklore, religious or secular, the tale of Meribah offers a lesson in social sciences, in particular sociolinguistics. Having evolved as a cognitive tool for facilitating instruction and toolmaking, language quickly became a tool for persuasion. As such, it also became a kind of force multiplier: in addition to being able to make and operate a stone ax and perhaps teach a few others to do so, one could now raise a posse, or an army, simply by choosing carefully what to say to whom, and when.

The next development — the fourth consciousness revolution, one that arguably caused human language to stand out from among all animal signaling and communication systems — was the transition from unidirectional to reciprocal persuasion, which serves cooperation and ultimately the forging of a consensus.[60] The new, peculiarly human social skills are about more than commanding the people (or magical water-spouting rocks) effectively. These skills, a litmus test for which the biblical Moses ultimately failed, include the ability to cultivate in people several special varieties of consciousness: a feeling of belonging together while respecting everyone's individuality; a recognition of shared destiny, which all can shape together; and a desire for achieving widespread agreement, — ideally, *consensus* — on what it is and how it should be shaped.[61]

> If the collective is indeed integral, questions are resolved not through majority and minority, but rather through unanimity, consensus.
> — Aleksandr Bogdanov (1920, p.77)

While language as we know it could not have evolved if it had no social functions, its availability enabled individuals to conceive of questions and ponder aspects of reality and of their own nature that would otherwise have remained forever beyond their ken. Language makes it possible for a person to acquire and nurture consciousness of processes that are too slow, too fast, too minute, or too vast to grasp directly through perception. One such process is the steady accumulation of weeks, months, and years that makes up a human life (which, according to John Lennon, is what happens to you while you are busy making other plans), as reflected in the mirror of the narrative self, or *soul*, that grows from the seed of mere sentience.[62]

> And the end of life is to fashion a soul, an immortal soul; a soul of one's own making. Because at the hour of death one bequeaths a skeleton to the earth, but a soul, a piece of work, to history.
> — Miguel de Unamuno (1974, p.6)

Another process that language helps reveal and comprehend is the shape-shifting tower of consciousness itself, in which the narrative is but one story. We can only reflect on our consciousness in this manner because language allows us to construct a description of it and take it all in. Thus, the entire tower — from its foundations in minimal consciousness, through effective selves, to the higher-order, reflexive consciousness or self-awareness — gets to be magically contained in one of its levels: the observatory chamber and stargazing platform at the tower's apex.

Language is the defining trait of our species, that which has set us on the course of taking over the planetary ecosystem, spelling ruin for all other species we share it with, and possibly dooming our own continued existence. At the same time, as I already noted, language is a distinct source of *hope*.[63] Indeed, the very concept of hope would not exist without language and its powers of reason and persuasion. While many living individuals have the power

> Reason cannot blossom without hope, and hope cannot speak without reason: both must operate in a Marxist unity; no other science has a future, no other future has science.
> — Ernst Bloch (1971, p.33)

to contribute more to the destruction of the world than all the generations past, the power to save it rests only with the society at large. It is therefore the relationship between the self and society that I turn to in the next chapter.

❧ ❧ ❧

Notes

1. Jorge Luis Borges, himself an adept in the magical uses of language, has remarked on the idea that language originated in magic in passing, in a public lecture he gave at Harvard in 1967 (Borges, 2000, p.81). The Marx quote in the epigraph is from his letter to A. Ruge, written in 1843 (Marx, 1992, p.209).

2. In arguing that many varieties of consciousness do not require language, I follow Churchland (1983). A key variety that does is social consciousness (Burns and Engdahl, 1998), as we'll see in the remainder of this book.

3. The case for the dependence of language on consciousness has been articulated in detail by Zlatev (2008).

4. Dissociation is defined in DSM-5 (the diagnostic and statistical manual of psychiatric disorders, version 5) as "disruption of and/or discontinuity in the normal integration of consciousness, memory, identity, emotion, perception, body representation, motor control, and behavior" (Lyssenko et al., 2018).

5. Is our world indeed "devoid of magic"? In what Bostrom (2003) calls the "basement reality" (as opposed to a simulated one, such as the virtual world of a computer game), magical acts that circumvent the laws of physics are not possible. In contrast, in a simulation, the architect of the virtual environment, or even just a regular agent who happens to know some cheat codes, can perform acts that can only be described as magical. Whether or not the world in which I wrote this book and you are reading it *is* the basement reality is a question that is not easily settled, as I have argued elsewhere (Edelman, 2011). Be that as it may, there are certain computational tricks that users of language can pull off that would be indistinguishable from magic, at least to a casual observer. For instance, having a detailed model of a target person (based on knowledge of their mindset and history) may enable someone else to manipulate them with great effectiveness. This may be the basis of folk beliefs in spells and witchcraft, found in all cultures, traditional (Tambiah, 1968) or "modern" (de Waal Malefyt, 2021; Subbotsky, 2014). Interestingly, evolution may have set us up for susceptibility to certain kinds of "spells," such as those that are cast by infants over their caregivers, on whom they depend for their very survival. Babies learn early on how to manipulate the people around them and cause them to perform various useful services, such as feeding, cleaning, or entertainment. For instance, Goldstein et al. (2009, p.636) report that by the age of 5 months, babies "have learned the [causal] social efficacy of their vocalizations on caregivers' behavior."

6. The words of Shimon bar Yochai are from the Babylonian Talmud, Berachot 35:2.

7. Here's a typical example of Chomsky's (1972, p.97) dismissal of evolution as it applies to language: "It is perfectly safe to attribute this development [of innate grammar] to 'natural selection', so long as we realize that there is no substance to this assertion." The formalist framework, which is still defended by some linguists (e.g., Bolhuis et al., 2014; Hauser et al., 2014), is being contested by the proponents of functionalist and usage-based approaches (e.g., Bates and MacWhinney, 1982; Diessel, 2015; Tomasello, 2003), according to which language is governed by the same domain-general computational principles as other behaviors, such as navigation (Edelman, 2017).

8. While philosophers have always been interested in the problem of consciousness, psychologists and neuroscientists tended to give it a pass (with some exceptions; e.g., Lashley, 1923; Natsoulas, 1981; Sperry, 1969). The proposal to focus on the neural correlates of consciousness instead of seeking after an exhaustive account, made by Francis Crick and Christof Koch (1990), effectively ended that tendency. The following years saw an explosive growth of the field of consciousness studies.

9. From general to specific, proposals regarding the context that could have facilitated language evolution include social intelligence (e.g., Worden, 1998), teaching (Szathmáry, 2015), and toolmaking instruction (Kolodny and Edelman, 2018; Stout, 2018).

10. In a paper titled *The origin of language in the Lower Palaeolithic*, Barham and Everett (2021) argue that "the origins of language can be detected one million years ago, if not earlier, in the archaeological record of *Homo erectus*."

11. Jakobson's (1960) well-known list of the six functions of language is representative of the contemporary literature. (A magical seventh function is the subject of a recent novel by Laurent Binet (2017).) In particular, J. L. Austin's (1962) book, *How to Do Things with Words*, focused on the performative and illocutory functions, and discussed, in addition, non-lexical content, such as mood, tone, cadence, emphasis, gestures, context (p.73ff), as well as "social behabitives" (p.159). In connection with the latter, Comoletti and Drout (2001, p.118) point out that "Speech acts or, to use Austin's original term, 'illocutionary acts,' are human utterances that affect the social world."

12. The effect of the pack size on the success rate of wolves hunting bison is reported in (MacNulty et al., 2014).

13. A commanding argument for the ubiquity of cooperation in non-human animals can be found in Pëtr Kropotkin's (1902) groundbreaking survey *Mutual Aid: A Factor of Evolution.*

14. Theories of language of the Construction Grammar (CxG) variety (e.g., Croft, 2001; Diessel, 2015; Goldberg, 2005; Tomasello, 2003) do not distinguish between lexicon (the store of words) and grammar (the system of rules). It's constructions all the way down: "As conceived in the present framework,

the grammar of a language is simply an inventory of linguistic units" (Langacker, 1987, p.63).

15. This conception of meaning is similar to that expressed by anthropologist Bronislaw Malinowski, as quoted by Tambiah (1968, p.186): "Meaning is the effect of words on human minds and bodies and through these on the environmental reality as created or conceived in a given culture" (Malinowski, B. (1965). *Coral gardens and their magic*, vol. I. Bloomington: Indiana University Press, p.53).

16. I subscribe to the following view of the ultimate function of language: "[W]e do not accept that language is primarily 'for' [...] communication, construed as exchanging coded meanings (as per the functionalist approach). Rather, we assume that language constitutes a toolkit that effectively supports an individual in influencing the state and the behaviour of others (and perhaps also self[)]" (Kolodny and Edelman, 2018, p.3).

17. An argument to the effect that writing can reshape cultural consciousness can be found in Eric Havelock's (1963) *Preface to Plato*: "Havelock portrays writing as inducing a restructuring of consciousness so profound that the very epistemology of a culture was altered by the technology: a change in technology induced a change in epistemology" (Dunn Neeley, 1998, p.203).

18. The quote (in my translation) is from Bakhtin (2002, p.100). In a section that discusses the dialogic nature of language, Di Paolo (2021, p.801) too cites Bakhtin (1984, p.293):

> The single adequate form for *verbally expressing* authentic human life is the *open-ended dialogue*. Life by its very nature is dialogic. To live means to participate in dialogue: to ask questions, to heed, to respond, to agree, and so forth. In this dialogue a person participates wholly and throughout his whole life: with his eyes, lips, hands, soul, spirit, with his whole body and deeds.

Freire (1974, p.40) similarly attaches critical importance to dialogue: "Born of a critical matrix, dialogue creates a critical attitude (Jaspers). It is nourished by love, humility, hope, faith, and trust. When the two "poles" of the dialogue are thus linked by love, hope, and mutual trust, they can join in a critical search for something. Only dialogue truly communicates." Freire's text continues with a quote from Jaspers:

> Dialogue is the only way, not only in the vital questions of the political order, but in all the expressions of our being. Only by virtue of faith, however, does dialogue have power and meaning: by faith in man and his possibilities, by the faith that I can only become truly myself when other men also become themselves.

In linguistics, a strong case for dialogic syntax has been made by Du Bois (2014). Friston and Frith (2015) cast communication as inference about the behaviour of others and self, thus placing it within the theoretical framework of predictive processing (discussed in this book at length in chapter 3). They further suggest that the problem of infinite regress that arises when two predictive agents attempt to model each other (which may be compared to the predicament of an agent that implements a Good Regulator; see p.68) can be "finessed" if both communicating agents possess the same model. Montemayor (2021, p.471) stresses the inherently social nature of the process of language use: "[...] representing the world in a way that makes conversational common ground salient is an essentially collective task that we can only achieve jointly with other speakers." He also discusses at length Turing's rejoinder to the "objection from consciousness" to his famous test for computer intelligence (Turing, 1950).

19. Concerning the voices of the dead, see (Edelman, 2008, section 9.6.2, "Ghosts in the machine").

20. Literally everything written by Le Guin and by Borges is essential reading as far as I am concerned. For some specific recommendations, see (Edelman, 2020).

21. The church father quoted by Borges (1999a) in this connection is St. Clement of Alexandria, who also wrote "The most prudent course is not to write but to learn and teach by word of mouth, because what is written remains" (*Stromata*).

22. The quote on reading as conscious action is from (Freire, 1985, p.19).

23. As Marshall Sahlins (1999, p.413) remarked, the tangle of culture, language, and consciousness may be hard to unravel: "Built into perception, endemic in the grammar, working in the habitus, [cultural] structure is the organization of conscious experience that is not itself consciously experienced. Not easy to figure."

24. For a review of the concept of collective consciousness in sociology, from Durkheim on, see (Wortmann, 2016).

25. The quoted passage is from (Havelock, 1963, p.141). In her review of Havelock, Dunn Neeley (1998, p.197) writes: "To position the Homeric consciousness as an oral culture, Havelock accounts for embedded public customs and private behavior codes — *nomoi* and *ethea* — as accepted conventions that when presented in epic narrative, are pedagogical and paradigmatic." The concept of collective consciousness is key to understanding the reasons for Plato's repudiation of epic poetry: "Havelock explains that Plato viewed the epic poet as teacher and that in his estimation, the act of individuals building a collective consciousness through active personal identification that required both physical and mental response was a morally dangerous and inappropriate way of learning, hence a threat to the future of Greek education" (Dunn Neeley, 1998, p.197).

26. See (Hanif, 2021, p.64): "The greatest epic poetry in Persian is *Shahnameh*, accomplished sometime around 1010AD. This text is replete with tales, epic narratives, romances, etc. in verse and has significantly affected the collective consciousness of the (nationalist) Iranians."

27. The editor of the Poona Critical Edition of the *Mahabharata*, V. S. Sukthankar, considers it to be "the content of [the Indian] collective unconscious... We must therefore grasp this great book with both hands and face it squarely. Then we shall recognize that it is our past which has prolonged itself into the present. We are it." (Quoted in *The Dilemma of a Liberal Hindu* by G. Das, in *Pluralism and Democracy in India: Debating the Hindu Right*, edited by Wendy Doniger and Martha C. Nussbaum, chapter 11, p.217.). Prasannan (2019, p.77), in quoting the same passage by Sukthankar, rather understandably changes the key phrase to "collective consciousness."

28. Hannah Arendt (2007a) ties the Jewish *haskalah* movement to the European Enlightenment — a difficult concept in its own right (see Edelman, 2020, ch.13,34,36,37, and of course the entire *Dialectic of Enlightenment* by Horkheimer and Adorno (2002), on why the Enlightenment was not actually as great a victory for humanism as it is usually portrayed).

29. See the entry "Borders and Boundaries" by Amilhat Szary (2015) in the *Handbook of Political Geography*. Goettlich (2019) describes the rise of linear borders in international politics.

30. For an in-depth analysis of the constitution of the Haudenosaunee Confederation, see (Lutz, 1998).

31. In a paper titled *How to do Other Things with Words*, Dennett (1997, p.234) writes:

> The acts and events you can tell us about, and the reasons for them, are yours because you made them — and because they made you. What you are is that agent whose life you can tell about.

Gallagher's (2000) review of the kinds of selves includes a section on the narrative self. For a psychological-developmental perspective, see *Autobiographical Memory and the Construction of a Narrative Self* by Fivush and Haden (2003). Richardson (1990) underscores the centrality of narrative in sociology.

32. In his essay *Los cuatro ciclos*, Borges suggests that there are just four basic plots: the siege of the city, the return home, the quest, and the sacrifice of God (*El Oro De Los Tigres*, pp.127-130, Buenos Aires: Emecé, 1972).

33. According to Polkinghorne (1991, p.135), the narrative self grows out of the common concepts and stories shared by the surrounding culture: "One's self-concept or self-identity is fashioned by adaptation of plots from one's cultural stock of stories and myths."

34. The description of the Homeric consciousness as "life without self-examination" appears in (Havelock, 1963, p.190). Here is the most relevant passage from (Havelock, 1963, pp.46-47) in its entirety:

> We must realise that works of genius, composed within the semi-oral tradition, though a source of magnificent pleasure to the modern reader of ancient Greek, constituted or represented a total state of mind which is not our mind and which was not Plato's mind;

> and that just as poetry itself, as long as it reigned supreme, constituted the chief obstacle to the achievement of effective prose, so there was a state of mind which we shall conveniently label the 'poetic' or 'Homeric' or 'oral' state of mind, which constituted the chief obstacle to scientific rationalism, to the use of analysis, to the classification of experience, to its rearrangement in sequence of cause and effect.

See also (Dunn Neeley, 1998). In a 1976 book, provocatively titled *The origin of consciousness in the breakdown of the bicameral mind*, Julian Jaynes (1976) advanced a similar idea, tying the "breakdown" to the catastrophic eruption, about 3,600 years ago, of the Thera volcano, which may have led to the collapse of the Minoan civilization and has certainly affected daily life, and politics, in the entire Eastern Mediterranean.

35. For a discussion of "the distinction between *self* (to begin with, the minimal self) and *person*," see (Crowell, 2021, p.265).

36. In a paper titled *Personal identity is social identity*, Carr (2021, p.341) describes the former as follows:

> It is an identity that is not pre-given a priori but is always in some sense an open question, never completely decided. Here the narrative conception of self is relevant, since it is often a question of what story or kind of story my identity instantiates. This notion of personal identity is inherently temporal, but not in the sense of temporal persistence but of temporal coherence of past, present and future. And here the question of personal identity is inevitably social, since it is largely a question of what group I identify myself with, what social role I take myself to embody.

37. Evidence for the coexistence of distinct selves in the different languages that one is fluent in can be found in (e.g., Caldwell-Harris, 2014; Schrauf and Rubin, 2003). It should be noted that Clarke (1976), who describes "second language acquisition as a clash of consciousness," goes way too far in equating bilingualism with "schizophrenia."

38. I shall return to the problem of false consciousness in chapter 7; meanwhile, see (Augoustinos, 1999; Jameson, 1981; Levy, 1991; Wilson, 2021). Du Bois (2018) had earlier discussed a related phenomenon of double consciousness. The role of language in shaping identity, in particular in the context of minoritization, has been examined, for instance, in (Fanon, 1967; hooks, 1994; Outlaw, 1974).

39. Pappé (2013) discusses "the rapid transformation of the Jews of Europe from the ultimate victims of that continent and that civilisation to the chief victimisers in Palestine." For a quick comparison between the decades-long Israeli occupation of Palestinian territories and other instances of foreign occupation, see (Bhan and Duschinski, 2020).

40. Hannah Arendt, whom Piterberg (2007) called "Zion's rebel daughter," noted this about Bernard Lazare (Arendt, 2007b, p.339):

> What he [Lazare] sought was not an escape from antisemitism but a mobilization of the people against its foes ... He did not look around for more or less antisemitic protectors but for real comrades-in-arms, whom he hoped to find among all the oppressed groups of contemporary Europe.

41. The Bundist ideology was based on the concept of *do'ikayt* — "here and now" — as noted by Brumberg (1999, p.207), who, unfortunately exhibits a lack of understanding of the socialist agenda and all but gloats over the Bund's loss to its adversaries, implying that the Zionist settlement in Palestine was the right way to go. In comparison, Kassow (2017, p.127) remarks that "the Bolshevik revolution and the Nazis destroyed the Bund, but the Nazis could easily have destroyed Zionism as well" (if Rommel had better luck in North Africa).

42. See *Concerning the Union of the Five Oppressed Groups, or the Pananarchy* (Gordin, 1919, p.105).

43. Here is the exhausted Gandalf telling his companions about his run-in with the Balrog, an entity that almost beat him at spells and wizardry Tolkien (1954, *The Fellowship of the Ring*, Book 2, Ch. 5):

> Then something came into the chamber — I felt it through the door, and the orcs themselves were afraid and fell silent. It laid hold of the iron ring, and then it perceived me and my spell.
>
> What it was I cannot guess, but I have never felt such a challenge. The counter-spell was terrible. It nearly broke me. For an instant the door left my control and began to open! I had to speak a word of Command.

The actual language of the Command is left unspecified by Tolkien.

44. Here is Roscigno (2011, p.355, emphasis in the original) on the faces of power:

> Bachrach and Baratz (1962) contend that a conception of power that only focuses on overt usage represents a limited one-dimensional view. It neglects a typically hidden, *second face of power wherein agendas are set, biases are mobilized and non-decision-making ensures that hierarchies remain intact* (see also Clegg 1989). Both Lukes (1974), in his seminal theoretical elaboration, and Gaventa (1982), in his case analyses of quiescence in an Appalachian community, articulate further *a third dimension and face of power wherein consciousness, interests, and beliefs are altered or manipulated in a manner that shapes compliance, conformity and passive agreement.*

45. See the review of *Policing: a Field Guide* by Shanahan (2019, p.140):

> A large part of *Police* consists of a critical interrogation of the Orwellian lexicon by which police murders become so-called "officer-involved shootings" and chokeholds "vascular restraints," violence of all sorts is rendered a "distraction technique," "Excited Delirium Syndrome" functions as a faux medical diagnosis for virtually anything a person does that results in their assault by a cop, and a "furtive movement is whatever a cop says it is" (18 ,93, 248, 259).

For the role of newspapers in the legitimation of police violence, see (Hirschfield and Simon, 2010).

46. The poem is included in (Jordan, 2005). See (Betts, 2012) for a comparative analysis.

47. "[A]t least forty percent of police officer families experience domestic violence as opposed to only ten percent of families generally" (Cheema, 2016, p.487). Regarding police crime in general, see (Boateng et al., 2021).

48. The defense spending data are from (Tian et al., 2020).

49. See (MacLeod and Chomsky, 2019) for an interview that offers a retrospective on *Manufacturing Consent* (Herman and Chomsky, 2002).

50. See (Edelman, 2020, ch.37), *War*. Kent (2019) argues that politicians who keep promising to end hunger are "not serious" about it. Scientists' analyses of the world hunger situation differ: some claim that enough food is already being produced for 10 billion people (Holt-Giménez et al., 2012), while others predict that needs will be met by 2050 (Berners-Lee et al., 2018). Hickel (2019) makes the critically important case for the possibility of achieving abundance while practicing degrowth.

51. A prominent example here is the Chinese Communist Party's language and "education" policies that contribute to the Uyghur genocide, presently under way in Xinjiang (Smith Finley, 2021). For information on China's enforcement of linguistic uniformity, see (Lin and Jackson, 2021; Spolsky, 2021).

52. The fragmentation of political discourse in the U.S. is often described as polarization (Levin et al., 2021). This is a poor choice of wording, because it implicitly assigns equal weight to "both sides" of the political spectrum — as if the spectrum is one-dimensional. Moreover, it is a dangerous choice, because it ignores that several of the many sides in contemporary political debate in this country openly embrace fascist ideology and practices.

53. An analysis of the historical trend in the use of political language conducted by Scheffer et al. (2021) has revealed that "over the past decades, there has been a marked shift in public interest from the collective to the individual, and from rationality toward emotion." The dynamics of post-truth discourse is discussed by Kalpokas (2019) and Ghosh (2022) in the context of politics, by Iyengar and Massey (2019) in science, and by Peters (2017) in education.

54. The quote is from (Marx, 1992, p.209).

55. This is Findlay (2009, p.27), quoting from the accompanying text to the *Manifesto*, which he edited (Marx and Engels, 2004).

56. The famous call for "ruthless criticism" is from (Marx, 1992, p.207, italics in the original).

57. As befits political anarchism, the title of Zube's (1977) book, *An Anarchist Manifesto*, uses the indefinite article, as compared to the definite article in the title of *The Communist Manifesto*.

58. See (Adorno and Horkheimer, 2010).

59. Helfgot (1993) observes the critical shift in God's statement of the means whereby water is to be extracted from the rock: whereas in Exodus 17 Moses is told to *strike* the rock, in Numbers 20 he is commanded to *speak* to it. It may be noted that Moses's speech impediment, which he had offered to the burning bush as an excuse for refusing his mission (Exodus 4:10), does not apply at Meriba, as he still has with him Aaron, his God-appointed spokesman.

60. Language is one of the "major evolutionary transitions" discussed by Szathmáry and Maynard Smith (1995), who link its evolution to the emergence of social cooperation at the level of the group. On a related note, Benítez-Burraco et al. (2021, p.2) write: "We contend that the characteristic traits of human language use are strictly connected to the transition from a mostly unidirectional form of persuasive communication (inherent both in human and animal communication) to a form of communication governed by persuasive reciprocity."

61. The importance of consensus in Iroquois politics is highlighted by Lutz (1998). The standard anarchist preference for consensus over majority rule is motivated by Bogdanov (1920, p.76) as follows:

> How do democrats resolve issues? By voting. What does voting mean? Counting individual units. As we can see, here individuals are held to be absolutely identical and equal. One may think this is something that is very good, but it actually reduces to making an abstract unit, a countable unit, out of a person; then the side where there are more units is recognized as the one that is in the right, while the side with fewer units is required to back down or withdraw: submit or leave. This is not collectivism, because submission is automatic.

Field studies such as that of (King et al., 2011) suggest that aiming at a consensus is not impractical: "We found that communicating groups (n=22) foraged more effectively than non-communicating groups (n=21) and were able to reach consensus decisions (an 'agreement' on the most profitable foraging resource) significantly more often than non-communicating groups" (p.1).

62. Translator's foreword to (de Unamuno, 1972, p.xv) quotes from his *Agony of Christianity*: "The finality of life is to make oneself a soul." This idea is echoed by Carr (2021), who adds (p.344): "Aristotle pointed out that misfortune late in life, like that of Priam, can still radically alter the tale. He suggested the paradox that one can count a person happy only at the end of her

life, when all the evidence is in, so to speak, or perhaps after she is dead, when the person is in no position to enjoy it. (See Nic. Ethics 1100a)" The notion of life as soul-making has, however, been best expressed by John Keats in a letter to his brother and sister-in-law:

> The common cognomen of this world among the misguided and superstitious is 'a vale of tears" from which we are to be redeemed by a certain arbitrary interposition of God and taken to Heaven — What a little circumscribe[d] straightened notion! Call the world if you Please "The vale of Soul-making" Then you will find out the use of the world...

(source: *The Letters of John Keats*, 1814-1821, edited by Hyder Edward Rollins (Harvard University Press, Cambridge, MA, 1958), pp. 100-104.)

63. Ernst Bloch's concept of *docta spes*, educated hope, is the central theme of his book, *The Hope Principle* (1971); see also (Levitas, 1990).

References

Adorno, T., & Horkheimer, M. (2010). Towards a new manifesto? *New Left Review, 65*, 33–61. Translated by Rodney Livingstone.

Amilhat Szary, A.-L. (2015). Boundaries and borders. In J. Agnew, A. Secor, J. Sharpe, & V. Mamadouh (Eds.), *Handbook of political geography* (pp. 13–25). New York: Wiley-Blackwell.

Arendt, H. (2007a). The enlightenment and the Jewish question. In J. Kohn & R. H. Feldman (Eds.), *Hannah Arendt: The Jewish writings* (pp. 79–94). New York, NY: Schocken Books. Originally published in 1942.

Arendt, H. (2007b). Herzl and Lazare. In J. Kohn & R. H. Feldman (Eds.), *Hannah Arendt: The Jewish writings* (pp. 338–342). New York, NY: Schocken Books. Originally published in 1942.

Augoustinos, M. (1999). Ideology, false consciousness and psychology. *Theory & Psychology, 9*(3), 295–313.

Austin, J. L. (1962). *How to do things with words*. Cambridge, MA: Harvard University Press.

Bakhtin, M. M. (1984). Appendix II: Toward a reworking of the Dostoevsky's book (1961). In *Problems of Dostoevsky's poetics*. Minneapolis, MN: University of Minnesota Press. Translated by C. Emerson.

Bakhtin, M. M. (2002). Problems of Dostoevsky's poetics. In *Bakhtin, M. M. collected works* (vol. 6). Moscow: Russian Dictionaries.

Barham, L., & Everett, D. (2021). Semiotics and the origin of language in the Lower Palaeolithic. *Journal of Archaeological Method and Theory, 28*, 535–579.

Bates, E., & MacWhinney, B. (1982). Functionalist approaches to grammar. In E. Wanner & L. Gleitman (Eds.), *Language acquisition: The state of the art*. New York, NY: Cambridge University Press.

Benítez-Burraco, A., Ferretti, F., & Progovac, L. (2021). Human self-domestication and the evolution of pragmatics. *Cognitive Science, 45*, e12987.

Berners-Lee, M., et al. (2018). Current global food production is sufficient to meet human nutritional needs in provided there is radical societal adaptation. *Elem Sci Anth, 6*, 52. https://doi.org/10.1525/elementa.310.

Betts, T. (2012). Everytime they kill a black boy…: Representations of police brutality against children in poems by Audre Lorde, Jayne Cortez, and June Jordan. *Obsidian: Literature in the African Diaspora, 13*(2), 69–85.

Bey, H. (1985). *Chaos: Broadsheets of ontological anarchism*. Anti-Copyright. https://theanarchistlibrary.org/

Bhan, M., & Duschinski, H. (2020). Occupations in context — the cultural logics of occupation, settler violence, and resistance. *Critique of Anthropology, 40*(3), 285–297.

Binet, L. (2017). *The 7th function of language*. New York, NY: Farrar, Straus and Giroux.

Bloch, E. (1971).*On Karl Marx*. New York, NY: Herder and Herder. Original edition: *Über Karl Marx*, Suhrkamp Verlag (1968), from *Das Prinzip Hoffnung*, 3 vols. (*passim*), Suhrkamp Verlag (1959), Frankfurt am Main. Translated by John Maxwell. https://thecharnelhouse.org/wp-content/uploads/2018/05/Ernst-Bloch-On-Karl-Marx.pdf

Boateng, F. D., Pryce, D. K., & Hsieh, M.-L. (2021). The criminal police officer: Understanding factors that predict police crime in the United States. *Crime & Delinquency*, https://doi.org/10.1177/00111287211054732

Bogdanov, A. (1920). *Elements of proletarian culture in the development of the working class* (in Russian). Moscow: Gosizdat. http://escriptorium.univer.kharkov.ua/handle/1237075002/1511

Bolhuis, J. J., Tattersall, I., Chomsky, N., & Berwick, R. C. (2014). How could language have evolved? *PLoS Biology, 12*, e1001934.

Borges, J. L. (1922/1999b). The nothingness of personality. In E. Weinberger (Ed.), *Selected non-fictions* (pp. 3–9). New York, NY: Viking.

Borges, J. L. (1999a). The cult of books. In E. Weinberger (Ed.), *Selected non-fictions* (pp. 358–362). New York, NY: Viking.

Borges, J. L. (2000). *This craft of verse: The Charles Eliot Norton lectures 1967–1968*. Cambirdge, MA: Harvard University Press. Edited by C.-A. Mihăilescu.

Bostrom, N. (2003). Are you living in a computer simulation? *Philosophical Quarterly, 53*, 243–255.

Brumberg, A. (1999). Anniversaries in conflict: On the centenary of the Jewish Socialist Labor Bund. *Jewish Social Studies, 5*(3), 196–217. New Series.

Burns, T. R., & Engdahl, E. (1998). The social construction of consciousness. Part 1: Collective consciousness and its socio-cultural foundations. *Journal of Consciousness Studies, 5*(1), 67–85.

Caldwell-Harris, C. L. (2014). Emotionality differences between a native and foreign language: Theoretical implications. *Frontiers in Psychology, 5*, 1055.

Carr, D. (2021). Personal identity is social identity. *Phenomenology and the Cognitive Sciences, 20*, 341–351.

Cheema, R. (2016). Black and blue bloods: Protecting police officer families from domestic violence. *Family Court Review, 54*(3), 487–500.

Chomsky, N. (1972). *Language and mind.* Cambridge: Cambridge University Press.

Churchland, P. S. (1983). Consciousness: The transmutation of a concept. *Pacific Philosophical Quarterly, 64*, 80–95.

Clarke, M. A. (1976). Second language acquisition as a clash of consciousness. *Language Learning, 26*(2), 377–390.

Comoletti, L. B., & M. Drout (2001). How they do things with words: Language, power, gender, and the priestly wizards of Ursula K. Le Guin's Earthsea books. *Children's Literature, 29*(1), 113–141.

Crick, F., & Koch, C. (1990). Towards a neurobiological theory of consciousness. *Seminars in the Neurosciences, 2*, 263–275.

Croft, W. (2001). *Radical construction grammar: Syntactic theory in typological perspective*. Oxford: Oxford University Press.

Crowell, S. (2021). On what matters. Personal identity as a phenomenological problem. *Phenomenology and the Cognitive Sciences, 20*, 261–279.

de Unamuno, M. (1972). *The tragic sense of life in men and nations.* Princeton, NJ: Princeton University Press. Translated by A. Kerrigan, with an Introduction by S. de Madariaga and an Afterword by W. Barrett.

de Unamuno, M. (1974). *The agony of Christianity. Selected works of Miguel de Unamuno* (vol. 5). Princeton, NJ: Princeton University Press. Translated by A. Kerrigan. Annotated by M. Nozick and A. Kerrigan.

de Waal Malefyt, T. (2021). Magical practice. In *Anthropology.* Oxford Research Encyclopedias. New York, NY: Oxford University Press.

Dennett, D. C. (1991). *Consciousness explained.* Boston, MA: Little, Brown & Company.

Dennett, D. C. (1997). How to do other things with words. *Royal Institute of Philosophy Supplements, 42*, 219–235.

Di Paolo, E. A. (2021). Enactive becoming. *Phenomenology and the Cognitive Sciences, 20*, 783–809.

Dickinson, S. (2018). *The monster Baru Cormorant*. New York: Tom Doherty Associates.

Diessel, H. (2015). Usage-based construction grammar. In E. Dąbrowska & D. Divjak (Eds.), *Handbook of cognitive linguistics*. Berlin: Mouton de Gruyter.

Du Bois, J. W. (2014). Towards a dialogic syntax. *Cognitive Linguistics, 25*, 359–410.

Du Bois, W. E. B. (2018). *The souls of black folk*. Musaicum Books. Originally published in 1903.

Dunn Neeley, S., Magee, D. B., & Enos, R. L. (1998). Review of *Preface to Plato* by Eric A. Havelock. *Rhetoric Review, 17*(1), 194–204.

Edelman, S. (2008). *Computing the mind: How the mind really works*. New York, NY: Oxford University Press.

Edelman, S. (2011). Regarding reality: Some consequences of two incapacities. *Frontiers in Theoretical and Philosophical Psychology, 2*, 44.

Edelman, S. (2017). Language and other complex behaviors: Unifying characteristics, computational models, neural mechanisms. *Language Sciences, 62*, 91–123.

Edelman, S. (2020). *Life, death, and other inconvenient truths*. Cambridge, MA: MIT Press.

Engels, F. (1949). *The part played by labour in the transition from ape to man. PRISM: Political & rights issues & social movements* (vol. 385). Moscow: Foreign Languages Publishing House. First published in *Die Neue Zeit* in 1876.

Fanon, F. (1967). *Black skin, white masks*. New York, NY: Grove Press. Translated by Charles L. Markmann.

Findlay, L. (2009). Karl Marx and Friedrich Engels, Manifest der kommunistischen Partei/The communist manifesto (1848). *Victorian Review, 35*(1), 23–27.

Fivush, R., & Haden, C. A. (2003). *Autobiographical memory and the construction of a narrative self: Developmental and cultural perspectives*. Mahwah, NJ: Erlbaum.

Freire, P. (1970). *Pedagogy of the oppressed*. New York: Continuum. English translation 2011.

Freire, P. (1974). *Education for critical consciousness*. New York: Continuum.

Freire, P. (1985). Reading the world and reading the word: An interview with Paulo Freire. *Language Arts, 62*(1), 15–21.

Friston, K., & C. Frith (2015). A duet for one. *Consciousness and Cognition, 36*, 390–405.

Gallagher, S. (2000). Philosophical conceptions of the self: Implications for cognitive science. *Trends in Cognitive Sciences, 4*, 14–21.

Ghosh, B. (2022). Politics of manufacturing consent in a post-truth society. *Journal of Developing Societies, 38*(1), 7–26. https://doi.org/10.1177/0169796X211068451

Goettlich, K. (2019). The rise of linear borders in world politics. *European Journal of International Relations, 25*(1), 203–228.

Goldberg, A. E. (2005). *Constructions at work: The nature of generalization in language*. Oxford: Oxford University Press.

Goldstein, M. H., Schwade, J. A., & Bornstein, M. H. (2009). The value of vocalizing: Five-month-old infants associate their own noncry vocalizations with responses from caregivers. *Child Development, 80*(3), 636–644.

Gordin, W. L. (1919). *Anarchist speeches* (in Russian.). Moscow: First Central Socialist State Technical College.

Hanif, M. (2021). The rise of the novel in Iran: A comparative study. *The Midwest Quarterly, 63*(1), 64–82.

Hauser, M. D., Yang, C., Berwick, R. C., Tattersall, I., Ryan, M. J., Watumull, J., Chomsky, N., & Lewontin, R. C. (2014). The mystery of language evolution. *Frontiers in Psychology, 5*, 401.

Havelock, E. A. (1963). *Preface to Plato*. Cambridge, MA: Belknap of Harvard University Press.

Helfgot, N. (1993). And Moses struck the rock: Numbers 20 and the leadership of Moses. *Tradition: A Journal of Orthodox Jewish Thought, 27*(3), 51–58.

Herman, E., & Chomsky, N. (2002). *Manufacturing consent: The political economy of the mass media*. New York, NY: Pantheon.

Hickel, J. (2019). Degrowth: A theory of radical abundance. *Real-World Economics Review, 87*, 54–68. Published 19 March 2019. http://www.paecon.net/PAEReview/issue87/Hickel87.pdf

Hirschfield, P. J., & Simon, D. (2010). Legitimating police violence: Newspaper narratives of deadly force. *Theoretical Criminology, 14*(2), 155–182.

Holt-Giménez, E., Shattuck, A., Altieri, M., Herren, H., & Gliefssman, S. (2012). We already grow enough food for 10 billion people ... and still can't end hunger. *Journal of Sustainable Agriculture, 36*(6), 595–598.

Hooks, B. (1994). *Teaching to transgress: Education as the practice of freedom*. New York, NY: Routledge.

Horkheimer, M., & Adorno, T. W. (2002). *Dialectic of enlightenment*. Stanford, CA: Stanford University Press. Originally published in 1944. Edited by Gunzelin Schmid Noerr. Translated by Edmund Jephcott.

Iyengar, S., & Massey, D. S. (2019). Scientific communication in a post-truth society. *Proceedings of the National Academy of Science, 116*, 7656–7661.

Jakobson, R. (1960). Linguistics and poetics. In T. Sebeok (Ed.), *Style in language* (pp. 350–359). Cambridge, MA: MIT Press.

Jameson, F. (1981). *The political unconscious: Narrative as a socially symbolic act*. Ithaca, NY: Cornell University Press.

Jaynes, J. (1976). *The origin of consciousness in the breakdown of the bicameral mind*. Boston, MA: Houghton Mifflin.

Jordan, J. (2005). Poem about police violence. In P. Townsend (Ed.), *Directed by desire: The collected poems of June Jordan* (pp. 272–273). Port Townsend: Copper Canyon Press.

Kalpokas, I. (2019). *A political theory of post-truth*. Cham: Palgrave Macmillan.

Kassow, S. (2017). The historiography of the Bund. *Polin Studies in Polish Jewry, 29*, 121–139.

Kent, G. (2019). GQ8 are we serious about ending hunger? *World Nutrition, 10*(3), 3–22.

King, A. J., Narraway, C., Hodgson, L., Weatherill, A., Sommer, V., & Sumner, S. (2011). Performance of human groups in social foraging: The role of communication in consensus decision making. *Biology Letters, 7*, 237–240.

Kolodny, O., & Edelman, S. (2018). The evolution of the capacity for language: The ecological context and adaptive value of a process of cognitive hijacking. *Philosophical Transactions of the Royal Society B, 373*, 20170052.

Kropotkin, P. A. (1902). *Mutual aid: A factor of evolution*. London: W. Heinemann.

Laing, R. D. (1967). *The politics of experience*. London: Penguin.

Langacker, R. W. (1987). *Foundations of cognitive grammar. Theoretical prerequisites* (vol. I). Stanford, CA: Stanford University Press.

Lashley, K. S. (1923). The behavioristic interpretation of consciousness. *Psychological Bulletin, 30*, Part I, 237–272; Part II, 329–353.

Le Guin, U. K. (1974). *The dispossessed*. New York, NY: Harper & Row.

Levin, S. A., Milner, H. V., & Perrings, C. (2021). The dynamics of political polarization. *Proceedings of the National Academy of Science, 118*(50), e2116950118. Introduction to Special Feature.

Levitas, R. (1990). Educated hope: Ernst Bloch on abstract and concrete utopia. *Utopian Studies, 1*(2), 13–26.

Levy, R. (1991). Structure-blindness: A non-ideological component of false consciousness. *International Journal of Sociology and Social Policy, 11*(6–8), 61–74.

Lin, C., & Jackson, L. (2021). Assimilation over protection: Rethinking Mandarin language assimilation in China. *Multicultural Education Review, 13*, 338–361.

Lutz, D. S. (1998). The Iroquois Confederation constitution: An analysis. Publius, 28(2), 99–127.

Lyssenko, L., Schmahl, C., Bockhacker, L., Vonderlin, R., Bohus, M., & Kleindienst, N. (2018). Dissociation in psychiatric disorders: A meta-analysis of studies using the dissociative experiences scale. *American Journal of Psychiatry, 175*, 37–46.

MacLeod, A., & Chomsky, N. (2019). Still manufacturing consent: An interview with Noam Chomsky. In A. MacLeod (Ed.), *Propaganda in the information age*. Chapter 1. Abingdon-on-Thames, UK: Routledge.

MacNulty, D. R., Tallian, A., Stahler, D. R., & Smith, D. W. (2014). Influence of group size on the success of wolves hunting bison. *PLoS One, 9*(11), e112884.

Marx, K. (1992). *Early writings*. New York, NY: Penguin. Introduced by Lucio Colletti. Translated by Rodney Livingstone and Gregor Benton.

Marx, K., & Engels, F. (2004). *The communist manifesto*. Peterborough, ON: Broadview. Originally published in 1848. Edited and translated by L. M. Findlay.

Montemayor, C. (2021). Language and intelligence. *Minds and Machines, 31*, 471–486.

Natsoulas, T. (1981). Basic problems of consciousness. *Journal of Personality and Social Psychology, 41*, 132–178.

Outlaw, L. (1974). Language and consciousness: Toward a hermeneutic of black culture. *Cultural Hermeneutics, 1*, 403–413.

Pappé, I. (2013). Shtetl colonialism: First and last impressions of indigeneity by colonised colonisers. *Settler Colonial Studies, 2*(1), 39–58.

Peters, M. A. (2017). Education in a post-truth world. *Educational Philosophy and Theory, 49*, 563–566.

Piterberg, G. (2007). Zion's rebel daughter: Hannah Arendt on Palestine and Jewish politics. *New Left Review, 48*, 39–48.

Polkinghorne, D. E. (1991). Narrative and self-concept. *Journal of Narrative and Life History, 1*(2–3), 135–153.

Prasannan, M. (2019). Traces of highbrowed innuendo from the epic: A kaleidoscopic reading of antedated scientific truths in the Mahabharata. *Research Lines, XII*(1–2), 77–81.

Richardson, L. (1990). Narrative and sociology. *Journal of Contemporary Ethnography, 19*(1), 116–135.

Roscigno, V. J. (2011). Power, revisited. *Social Forces, 90*, 349–374.

Sahlins, M. (1999). Two or three things that I know about culture. *Journal of the Royal Anthropological Institute, 5*(3), 399–421.

Scheffer, M., van de Leemput, I., Weinans, E., & Bollen, J. (2021). The rise and fall of rationality in language. *Proceedings of the National Academy of Science, 118*, e2107848118.

Schrauf, R., & Rubin, D. (2003). On the bilingual's two sets of memories. In R. Fivush & C. Haden (Eds.), *Autobiographical memory and the construction of a narrative self* (pp. 121–145). Mahwah, NJ: Erlbaum.

Shanahan, J. (2019). Review of "policing: A field guide" by David Correia and Tyler Wall (2018). *Social Justice, 46*(4), 139–143.

Smith Finley, J. (2021). Why scholars and activists increasingly fear a Uyghur genocide in Xinjiang. *Journal of Genocide Research, 23*(3), 348–370.

Sperry, R. W. (1969). A modified concept of consciousness. *Psychological Review, 76*, 532–536.

Spolsky, B. (2021). Qinsheng Zhou: Ethnic minority languages in China: Policy and practice (de Gruyter, 2020). *Language Policy, 20*, 291–294. Book review.

Stout, D. (2018). Archaeology and the evolutionary neuroscience of language: The technological pedagogy hypothesis. *Interaction Studies, 19*, 256–271.

Subbotsky, E. (2014). The belief in magic in the age of science. *SAGE Open, 4*(1). https://doi.org/10.1177/2158244014521433

Szathmáry, E. (2015). Toward major evolutionary transitions theory 2.0. *Proceedings of the National Academy of Science, 112*, 10104–10111.

Szathmáry, E., & Maynard Smith, J. (1995). The major evolutionary transitions. *Nature, 374*, 227–232.

Tambiah, S. J. (1968). The magical power of words. *Man, 3*(2), 175–208. New Series.

Tian, N., Kuimova, A., Lopes da Silva, D., Wezeman, P. D., & Wezeman, S. T. (2020). Trends in world military expenditure, 2019. Technical report, Stockholm International Peace Research Institute.

Tolkien, J. R. R. (1954). *The Lord of the Rings*. London: George Allen & Unwin.

Tomasello, M. (2003). *Constructing a language: A usage-based theory of language acquisition*. Cambridge, MA: Harvard University Press.

Turing, A. M. (1950). Computing machinery and intelligence. *Mind, 59*, 433–460.

Wilson, L. (2021). Why they know not what they do: A social constructionist approach to the explanatory problem of false consciousness. *Journal of Social Ontology, 7*(1), 45–72.

Worden, R. (1998). The evolution of language from social intelligence. In J. R. Hurford, M. Studdert-Kennedy, & C. Knight (Eds.), *Approaches to the evolution of language* (pp. 148–166). Cambridge, UK: Cambridge University Press.

Wortmann, S. (2016). Collective consciousness. In G. Ritzer (Ed.), *The Blackwell encyclopedia of sociology*. New York, NY: Wiley.

Zlatev, J. (2008). The dependence of language on consciousness. *Journal of Consciousness Studies, 15*(6), 34–62.

Zube, K. H. (1977). *An anarchist manifesto: The manifesto of peace and freedom (The alternative to the communist manifesto)*. Freiburg: Mackay-Gesellschaft. Translated from the German by Doris Pfaff and John Zube. Edited by Edward Mornin.

5 Self and Society

A beast or a god. The discovery of critical consciousness. Learning freedom. Selves and collectives. Morality and cooperation. Conscience and consciousness. The fifth revolution. The counter-revolution, or: What Went Wrong.

> At fifteen I set my heart on learning. At thirty I could stand. At forty I had no doubts. At fifty I knew the Decree of Heaven. At sixty I was already obedient [to this Decree]. At seventy I could follow the desires of my mind without overstepping the boundaries [of what is right].
>
> *Analects*
> — CONFUCIUS

> I scarcely remember a commandment, nothing appears to me in the form of a law, it is an instinct that guides me and always guides me; I follow my dispositions with freedom and know as little of limitation as of repentance.
>
> *Confessions of a Beautiful Soul*
> — GOETHE

> Insofar as the question [of the direction of history] comes up at all, it's usually when reflecting on why the world seems to be in such a mess and why human beings so often treat each other badly — the reasons for war, greed, exploitation, systematic indifference to others' suffering. Were we always like that, or did something, at some point, go terribly wrong?
>
> *The Dawn of Everything: A New History of Humanity*
> — D. GRAEBER & D. WENGROW[1]

A Beast or a God

I CREDIT my eventual becoming aware of the scope of my miseducation, on a cluster of subjects surrounding the relationship between the self and society, to a conjunction of two factors: a succession of solo hikes in the desert and a series of senior seminars that I taught over the years.

It's not that my upbringing and schooling had been ineffective. As a teenager, I had set my eyes on doing science, and that's what I did; most everything else was just a distraction.

> The work is done,' grown old he thought,
> 'According to my boyish plan
> Let the fools rage, I swerved in naught,
> Something to perfection brought'

S. Edelman, *The Consciousness Revolutions*,
https://doi.org/10.1007/978-3-031-24012-6_5

Gradually, things changed — or rather, I did. The "boyish plan" and the schooled persona that had been supervising its execution were no longer perceived as appealing by the newly aware *me*.[2]

There is nothing extraordinary about being in this state: those who don't wish they had done some things differently when they were young are either very lucky or not very discerning. But my concern here is not with regret, central as it may be to the condition of any self-conscious agent with a sense of personal history.[3] Instead, I want to invoke a bit of history, personal and other, to focus, first, on the process whereby the self becomes conscious of its place in society and, second, on the ensuing conception of agency, and of freedom.

I had been brought up to conceive of the relationship between the self and society in no uncertain terms: the unspoken assumption at home has always been that the former should strive for absolute freedom and independence from the latter. Growing up in a minoritized intelligentsia family in a pretend-communist totalitarian state, the 1960s and 1970s USSR, provided for two distinct ways in which such extreme individualist ethos could be reinforced. At home, the deep disdain for the lying regime, with its constant public exhortations for collectivism and communism, automatically predisposed one to their presumed antithesis, individualistic capitalism.[4] At school, the official denouncements of the "decadent and money-worshipping" West served as an effective endorsement.

Surely, we thought, the Soviet state propaganda, always peddling falsehoods about *our* prosperity and happiness, was also lying in its stories about how in the West people were brainwashed into worshiping the dollar and hating communism, while millions were unemployed, often hungry, unhoused, deprived of healthcare, lonely, and depressed. We had the same attitude to the official Soviet reports on the alleged racism of the U.S. political system. During the highly publicized 1972 trial of the Black activist Angela Davis, our entire class was made to write letters of support for her. If there was one thing that could put us off the idea of systemic injustice in America, it was being forced by our hypocrite Party handlers to decry it.

Свобода — это когда забываешь отчество у тирана
(Freedom is when you forget the tyrant's patronymic)
— Joseph Brodsky (1991)

Along with hundreds of thousands of other Jews, we were allowed to leave the Soviet Union about the same time the authorities sent into exile the poet, Nobel laureate, and famous dissident Joseph Brodsky. Like most of the people in my circle at the time, I equated freedom with being away from the oppressive Soviet regime; and like many emigrants, I never looked back once I got away: out of sight, out of mind.

It would be decades before I gave the question of freedom any serious thought. In the meanwhile, the world was changing. My old reference points for freedom politics went away with the disappearance of the Soviet empire, and new ones were being acquired as I learned first-hand about life in "the West." One thing remained constant, though: my privilege of being securely employed and white in a country where having enough money and the right skin color effectively insulates one from most of the "slings and arrows of outrageous fortune." This ensured that I did not *have* to think too hard about freedom, or politics, or the economy — indeed, about anything at all, except my science and my immediate family. Being, or at least being well-off, does after all determine consciousness.

Ironically, it was because of my privilege that I was able to take the first steps on the road that eventually brought me to where I stand now on freedom and consciousness. Our boys were off to college, and it occurred to me that I could use the spring and fall breaks from my teaching (which had long settled into a routine, tolerated but not enjoyed as much as science) to hike some not particularly family-friendly places that I had nevertheless always wanted to see, such as Death Valley. I could now well afford to fly to those places, even just for a week at a time, and so I did.

Being away from all human company for days at an end is a powerful experience, especially if one goes through it in one of the least hospitable-looking places on the planet. I loved it — but the solitude often seemed flawed.[5] Unless you go off into the desert for good (as true hermits do), and perhaps even then, there is a tether that connects you back to the human world, or at the very least to the hamster wheel of your routine worries and ruminations. I felt it pull on me, even as I was standing at the summit of a desert peak with a hundred-mile vista in every direction, or pitching my tent in a canyon with nobody else within a day's walking distance.

But he who is unable to live in society, or who has no need because he is sufficient for himself, must be either a beast or a god: he is no part of a state. — Aristotle (1885, I.2).

Note that "state" here means the community, or *polis* — not "the instrument of exploitation of the oppressed class" (Lenin, 1917).

After a series of such experiences, I decided to try and get to the bottom of it, using such tools as I had at my disposal: reading science, doing some, and writing about it. Why could I not be happy and free while standing on that summit? Why the constant consciousness of the moment's fleetingness and of the call — not of the wild, but of the "real world," from which I still had a few days' worth of freedom to enjoy? The answer, such as I was able to formulate and justify, took the form of a book, whose title pretty much encapsulated its main idea: *The Happiness of Pursuit*.[6]

The message of that book was upbeat: even though for evolutionary reasons happiness is necessarily transient, thus necessitating repeated pursuit ("catch and release" was how I put it), the same evolutionary processes put happiness within reach of everyone. My flirt with positive psychology — unwitting, as I don't believe I gave a second's worth of consideration to this inane phrase while working on that book — proved to be brief. After fielding a few poignant questions while being interviewed about happiness, I decided I should learn something about the context in which it plays out, or fails to do so, in practice. For that, I had to consider not just our innate capacity for happiness, but also the conditions that promote or impede it.

The Discovery of Critical Consciousness

A convenient framework for doing due diligence on the conditions for happiness (again, better late than never) was the "Topics" seminar that I had been teaching every semester for years. (Another sign of privilege: both the choice of topic and the composition of the syllabus for those seminars were entirely up to me.) The very next year after the happiness book came out, I ditched my usual rotation of topics (*Mind and Reality in Science Fiction*, *Borges and I*, and other such stuff) in favor of something more practical and arguably more important: *Inequality, Power, and Happiness*.

My usual recipe for getting up to speed in a new domain that I am trying to learn about is, first, to rely on primary literature instead of textbooks (even if such can be found); second, to get the students interested in the questions I put forward and have them ask questions of their own; and, third, to organize the seminar around free-for-all discussions. Four semesters of running my inequality seminar this way did help me learn some things about the political economy of happiness and freedom — and about propaganda. As someone who had lived in (and then outlived) the old USSR, I had prided myself on being immune to propaganda, only to discover that my *critical consciousness* was in need of an overhaul.

Critical consciousness is integrated with reality; naïve consciousness superimposes itself on reality; and fanatical consciousness, whose pathological naïveté leads to the irrational, adapts to reality.
— Paulo Freire (1974, p.39)

It must have been my automatic rejection of the horror stories peddled by the Soviets about life under capitalism, which I deemed to be mere propaganda, that had blinded me to the possibility that the West's narrative itself might be just that. It was hard data papers that we read and discussed, not theory or opinion pieces, that gradually helped me see that the list of the West's vaunted freedoms was, and still is, topped by the freedom to starve, to be unhoused, to go without medical care, all in the name of the empty promise of unlimited opportunity and the conceit of the redeeming power of hard work, extolled by a corporate-owned "free press" which serves the interests of those who themselves work the least.[7]

Bourgeois freedoms are always more bourgeois than they are freedoms.
— Ernst Bloch (1971, p.47)

It is easy to identify at least two of the causes behind my original failure to appreciate the complexity of the concept of freedom: one systemic and the other personal. The systemic cause I already mentioned: the privileged status conferred on me by my socioeconomic environment. Never in my life have I been in danger of going hungry, or becoming homeless, or being sick and unable to get help. Moreover, thanks to my parents' dedication, which included going to the great trouble of pulling up roots and moving to another country in pursuit of a better life, I had been presented with a series of excellent and free educational opportunities.

Unfortunately, education by itself, even if it does not hew to a state-dictated curriculum, is unlikely to foster free-thinking — unless one is lucky enough to have someone like Paulo Freire[8] or bell hooks for a teacher, or astute enough to overcome the limitations imposed by one's upbringing and social class. This brings me to the second, personal cause of my early obliviousness on questions of freedom: I was neither smart enough to cross the boundaries of inquiry established by my education, nor compassionate enough to at least pay attention to the world around me.[9]

It would indeed be naive to expect the oppressor elites to carry out a liberating education.
— Paulo Freire (1970, 71).

Learning and teaching about inequality helped me focus my attention and made me interested in the economic and psychological aspects of freedom — not the abstract freedom of the will, but rather the conscious experience of freedom in practical interactions between the self and others. This concept of freedom sits at the interface between the science of consciousness and political economics. For one thing, *class consciousness* is necessarily a collective or social phenomenon and at the same time an economic one. But even if we focus on those aspects of self-consciousness that have already been discussed so far in this book, a connection to sociology and economics is clear: the praxis of freedom is part of conscious experience insofar as it involves socially constrained perspective (on life), agency (over actions, such as are available to the self that is situated in a society), and aspects of embodiment (having to do with the procurement of food and shelter, as well as bodily autonomy).

We should not think of human agency in terms of free will, but rather look for a holistic notion of human freedom.
— Chenyang Li (2014, p.905).

Learning Freedom

All this called for another seminar, so I put one together, under the title *Varieties of Freedom*. Having promised my students variety, I made an effort to avoid both geopolitical parochialism and disciplinary myopia, to the extent possible within the limited time frame of one semester. This is why, in addition to some Western sources, we read papers by Chinese and African scholars, in disciplines that included philosophy, evolution and psychology, history and geography, anthropology, politics, and education.[10]

Of course, I had already been reading widely for a long time before taking on the theme of that seminar (and this book). What really helped me learn this time around was the range of perspectives that the students brought to the seminar. Every time a participant preceded the thoughts they were about to share with a self-identification as someone different from me in some respect — a work-study student, a woman, a person of color, a first-generation scholar, an immigrant from a place I knew little about — it felt as if I was given a gift, a glimpse of the human world through a window that would otherwise have remained shut.

Our expedition yielded immediate and ample rewards: the readings and the discussions helped us realize the degree to which the ultra-individualistic Western conception of the self and the corresponding notion of freedom as total absence of constraint are outliers, when seen in the proper historical and cross-cultural perspective. Even more immediately satisfying were the occasional discoveries we kept making in the literature of concise, yet effective arguments against the idea that rugged individualism is the proper practical philosophy of life, such as the observation that the variety of freedom that can be found in the wilderness is not necessarily a desirable one.

More to the point, a self-exile who seeks freedom in the desert must be exceptionally well-equipped to begin with, as well as periodically resupplied, to afford to spend any significant time in contemplation, as opposed to spending all available time in eking out a bare existence and likely failing at that. It's not that survival as such is infeasible. It is definitely possible for a human to survive even in the least hospitable places on earth — just not alone.[11] The native inhabitants of Death Valley, the Timbisha Shoshone, have been living there for a thousand years — collectively, in community. In contrast, not one of the handful of Anglo loners who settled in Death Valley to farm actually subsisted on self-produced food (and of course those who sought or worked mining claims never intended to be self-sufficient in the first place).[12] In the desert, it takes a village simply to *be*.

While providing an element of possibility for the agent, abstract freedom is not of supreme value; otherwise an unconstrained person lost in a boundless desert would have the utmost freedom.
— Chenyang Li (2014, p.910)

What of being free, then? As Hannah Arendt observed in a 1967 essay *The Freedom to Be Free*,

> Only those who know freedom from want can appreciate fully the meaning of freedom from fear, and only those who are free from both want and fear are in a position to conceive a passion for public freedom, to develop within themselves that *goût* or taste for *liberté* and the peculiar taste for *égalité* or equality that *liberté* carries within it.[13]

The inescapable subordination of personal freedom to community dynamics, underscored by the challenge of surviving in the desert, is equally prominent at the other extreme of the spectrum of human habitats — the "urban jungle," a phrase that connotes an entire series of archetypal images. The first, visceral one arises from the our primal fear of the jungle as a place of mortal danger. The second, romantic one pivots to viewing the jungle as a space for freedom, of the kind afforded to the well-to-do by slumming. The third, socioeconomic one closes the circle: jungle is now a metaphor, invoked by the pearl-clutching classes to refer to those parts of the city where they dread to tread.[14]

This, therefore, is the *dialectic of freedom*: personal freedom and, indeed, personal survival are conditioned on certain patterns of relationships between the individual and the community. One such pattern is that of exploitation. It is particularly conspicuous under capitalism, in all its incarnations: the classical, openly predatory variety castigated by Marx and revived in the latter-day United States and Britain; the "compassionate" kind of northern Europe or New Zealand; the state capitalism of the old USSR, where the poor, while politically unfree, had at least the benefits of steady employment, free health care, and guaranteed pension; and the "People's Republic" of China, where the hundreds of millions of poor have nothing but their chains.

The DIALECTIC OF FREEDOM: personal freedom and personal survival are conditioned on certain patterns of relationships between the individual and the community.

Under the exploitation model, those who have the means for it — from industry barons and corporate oligarchs to mere suburbanites and dwellers in gated communities — have more freedom than the denizens of "housing projects" and the "inner city" *because* of the work done by the latter. Both sides of this social contract hold on to the same twisted conception of freedom that values above all extreme individualism and "economic opportunity" — an astonishing consensus, given the tragic personal and societal price that is being paid for individualism and the vanishing probability of the realization of the alleged opportunity on part of the unfree masses.

This consensus epitomizes *false consciousness* that has been elevated to the status of a global secular religion, one with which the assorted official religions coexist more readily than they do with one another. It is, however, founded on a historical and evolutionary narrative that is deeply fallacious, and as such is vulnerable to being exposed, contested, and perhaps replaced by one that can truly deliver freedom for all at the expense of none.

Selves and Collectives

At the core of the cult of individualism are two related Bad Ideas: first, that *H. sapiens* is by nature predominantly egotistical and aggressive and, second, that only with the emergence of the political apparatus of the state was that wild nature tamed and peace and prosperity made possible. Both these ideas are familiar to us from textbooks. First, we are taught that because all life is an unending "struggle for existence," our human nature predisposes us to a war of all against all. The second idea then follows naturally: historical progress cannot be anything else but humanity's abandonment of life in the "state of nature" that is "red in tooth and claw" in favor of a series of modes of coexistence, in all of which people's innate tendency to conflict and violence is countered by concentrating political power in some form of a centralized *state*.[15]

Not surprisingly, state-sponsored textbooks rarely, if at all, discuss the lessons that can be learned from the global experiment in entrusting our well-being, and ceding our freedoms, to the state. It is, however, not difficult for an open-minded student of history, anthropology, or sociology to figure out how things have been going down in that regard.[16] Even before the 20th-century imperial and nation-state politics brought us two world wars and innumerable lower-key armed conflicts, all accompanied by record internal repressions and socioeconomic inequality, the track record of the institutions of state in safeguarding and promoting *all* of their citizens' well-being — and not just that of the state's minions, the *de facto* wielders of state power — has been about as good as that of a fox who arranges to be put in charge of a hen-house.

The response to this realization has been the emergence of principled political *anarchism*, for which Colin Ward offers the following brief introduction:[17]

> The anarchists and their precursors were unique on the political Left in affirming that workers and peasants, grasping the chance that arose to bring an end to centuries of exploitation and tyranny, were inevitably betrayed by the new class of politicians, whose first priority was to reestablish a centralized state power. After every revolutionary uprising, usually won at a heavy cost for ordinary populations, the new rulers had no hesitation in applying violence and terror, a secret police, and a professional army to maintain their control. For anarchists the state itself is the enemy, and they have applied the same interpretation to the outcome of every revolution of the 19th and 20th centuries. This is not merely because every state keeps a watchful and sometimes punitive eye on its dissidents, but because every state protects the privileges of the powerful.

The anarchist project of doing away with the mechanisms of state oppression by dismantling the state has always been countered with the argument that the newly liberated people would revert to the "state of nature" — a cost that few would deem acceptable. As political anarchism, which roundly rejected the Bad Idea that the state can ever be the people's best keeper, was taking shape in late 19th century, even its leaders were under the spell of the other Bad Idea: that human beings are by nature bad. Thus, Mikhail Bakunin wrote that man must "emancipate himself from the yoke of his own nature" to achieve a new, collective consciousness through education and training.

Advances in evolution science, anthropology, and history in the century and a half since then have made the project of educating the masses considerably easier, by exposing the two Bad Ideas as not just normatively deplorable, but as factually wrong. What we need to be educated about, over and above the perils of individualism and the moral importance of cooperation, is the centrality of the latter in our own evolutionary and historical legacy.

Already in 1905, geographer and anarchist Jacques Élisée Reclus, a contemporary of Bakunin and of Darwin, defended evolution science against the "social Darwinists" who ignored its findings and distorted its message:[18]

> When Darwin, Wallace and their emulators had so admirably laid out the system of organic evolution through adaptation of beings to their environment, most of their disciples ... allowed themselves to be seduced by a simplistic hypothesis, seeing only the 'struggle for existence' in the infinite drama of the living world. However, the renowned author of *On the Origin of Species* and *The Descent of Man* had also addressed the 'agreement for existence'. ... Yet how many alleged Darwinists wanted to completely ignore all the facts of mutual aid and angrily scream and shout, as if the sight of blood excited them to murder: 'The animal world is a gladiators' arena ... every creature is trained for

The human being becomes conscious of himself and his humanity only in society and only by the collective action of the whole society. ... He can emancipate himself from the yoke of his own nature, i.e. subordinate his instincts and the movements of his body to the conscious direction of his mind, the development of which is fostered only by education and training. But education and training are preeminently and exclusively social ... hence the isolated individual cannot possibly become conscious of his freedom.
— Mikhail Bakunin (1871)

> combat!' And under the cloak of science, how many violent and cruel people found themselves suddenly justified in their acts of selfish appropriation and brutal conquest!

An extensive collection of field observations in support of this view of evolution was published in 1902 by another geographer anarchist, Pëtr Kropotkin, under the title *Mutual Aid: A Factor of Evolution.*[19] The ascendance of social Darwinism, heavily promoted by those with a vested interest in its perceived justification of political and socioeconomic inequality, soon caused the work of Kropotkin, Reclus, and other scientific rebels to be relegated to the status of a quaint but largely irrelevant footnote to the development of evolutionary theory and its application to human sciences and history.

The eventual vindication of the mutualists had to wait until the turn of the 21st century. It was assisted by the emergence and consolidation of the Extended Evolutionary Synthesis, which afforded insights into multi-level selection and the co-evolution of genes and culture. Theoretical advances and computational simulations yielded clear evidence of the evolutionary feasibility of altruism and *cooperation*, defined as "costly behaviour performed by one individual that increases the payoff of others" (more about which later in this chapter). At the same time, new findings in ethology, archaeology, anthropology, sociology, and history have led to a renewed appreciation of the evolutionary, ancestral, historical, and contemporary importance of cooperation in the functioning of all known societies.[20]

Strikingly, this latter historical insight can be found already in Kropotkin's *Mutual Aid*:[21]

> [It is] utterly false to represent primitive man as a disorderly agglomeration of individuals, who only obey their individual passions, and take advantage of their personal force and cunningness against all other representatives of the species. Unbridled individualism is a modern growth, but it is not a characteristic of primitive mankind.

Just how prescient Kropotkin was in this matter can now be seen from the newly available extensive anthropological and archaeological evidence, collected, reviewed, and analyzed by David Graeber and David Wengrow and reported in their 2021 book, *The Dawn of Everything*. I will allow myself here just one long quote from that book, justified by its focus on consciousness:[22]

This corporate existence among those of his own species ceases, as far as the individual man is concerned, only with death, or, as the Latin so aptly puts it: *inter homines esse* to denote life, and *desinere inter homines esse* to denote death. Only in death is human existence completely and utterly individual.
— Hannah Arendt (1961a, p.205)

> Human thought is inherently dialogic. Ancient philosophers tended to be keenly aware of all this: that's why, whether they were in China, India or Greece, they tended to write their books in the form of dialogues. Humans were only fully self-conscious when arguing with one another, trying to sway each other's views, or working out a common problem. True individual self-consciousness, meanwhile, was imagined as something that a few wise sages could perhaps achieve through long study, exercise, discipline and meditation. What we'd now call political consciousness was always assumed to come first.
>
> In this sense, the Western philosophical tradition has taken a rather unusual direction over the last few centuries. Around the

> same time as it abandoned dialogue as its typical mode of writing, it also began imagining the isolated, rational, self-conscious individual not as a rare achievement, something typically accomplished — if at all — after literally years of living isolated in a cave or monastic cell, or on top of a pillar in a desert somewhere, but as the normal default state of human beings anywhere. Even stranger, over the course of the eighteenth and nineteenth centuries it was *political* self-consciousness that European philosophers came to see as some kind of amazing historical achievement: as a phenomenon which only really became possible with the Enlightenment itself, and the subsequent American and French Revolutions.

We do not have to subscribe to the strong thesis that holds even individual consciousness to be socially constructed[23] to appreciate the essential social engagement of the human sense of self. It is the dynamics of the interplay between individual and collective consciousness that gives rise to our explicit sense of morality — the enabler and preserver of our coexistence.

Morality and Cooperation

Suppose you sit down some dogs in front of a computer and have them choose repeatedly which displayed symbol to boop — one that carries a later reward for another dog (but not for themselves), or an unrewarded one. Now have some wolves do the same task. Which of the two species would you predict to be more prosocial?

My only encounter with wolves in the wild — at first light, on a solo hike in the red rock desert country north-west of Eilat — was not a very close one. Even under the 10x magnification of my telephoto lens, the two animals looked nondescript on the camera screen and could only be identified as wolves later, with the help of a zoologist friend. A few years later, I first saw a wolf at close quarters, in the little research zoo at Tel Aviv University (where my friend was a student at the time). The experience of looking into the wolf's yellow eyes was unforgettable and literally hair-raising: the atavistic reflex of my hominid ancestors must have kicked in.

The likely ancestral distrust of wolves, Little Red Riding Hood's widely publicized traumatic, but fictional experience, and classical proverbs such as *homo homini lupus est*, "man is wolf to man" — all these predispose people to expect a win for dogs in the prosociality match-up. Science, however, tells a different story. Behavioral experiments have shown that wolves, but not dogs, are prosocial in the touch screen task and outperform dogs in a test of cooperation with conspecifics.[24] That dogs are friendly to *us* is not to their credit: apparently, by domesticating dogs, we turned them into class traitors, more loyal to their masters than to their comrades.

Are humans in fact as poorly predisposed towards each other as the proverbial wolves? It helps to know that the familiar phrasing of the damning proverb is incomplete. In the earliest recorded source, which dates back to republican Rome, it comes with an important qualification: "A man is a wolf rather than a man to another man, *before he gets to know him*."[25] The upshot of this twist in the plot, when combined with Seneca's rejoinder to the wolf slur, *homo, sacra res homini* or "man, a sacred thing to man" (for the "discovery" of both of which I am grateful to Wikipedia, a free cooperative project), is that people are *conditional cooperators*.

The new evolutionary science of morality concurs. As noted already by Darwin (albeit not in this many words, because the biological basis of inheritance was not known in his time), genetic predisposition to altruism is not a stable trait, because altruists or, more generally, unconditional cooperators, are at a disadvantage relative to selfish individuals. Darwin's proposed explanation for the undeniable fact that *individuals* nevertheless often do cooperate with each other invoked mutual aid at a *group* level. According to this hypothesis, groups whose members tend to cooperate with each other (but presumably not with outsiders) outperform groups of selfish individuals. Darwin also suggested how *cultural evolution* (to use a modern term) might gradually raise the level of morality, insofar as it is expressed as *cooperation*, that is, costly behavior that benefits others.[26]

Selfishness beats altruism within groups. Altruistic groups beat selfish groups. Everything else is commentary.
— D. S. Wilson & E. O. Wilson (2007, p.345)

The Extended Evolutionary Synthesis, whose foundations include both multi-level selection and cultural evolution, has substantiated Darwin's insights. Perhaps unexpectedly, it also served to reconnect evolutionary theory to anthropology and history; as D. S. and E. O. Wilson remarked in discussing multi-level selection, "These ideas can potentially explain the broad sweep of recorded history in addition to the remote past."[27] Moreover it did so in a manner anticipated neither by the social Darwinists nor by those Marxists who were overly enthusiastic about the idea that "being determines consciousness." The former misjudged — or, if acting in bad faith, misrepresented — human nature by positing it to be "red in tooth and claw"; yet it emerged that humans are, after all, very much into mutual aid. In contrast, Marxist social engineers tended to deny the biological aspects of human nature altogether; yet their projects may have stood a better chance of success, had they taken into account both the natural human tendency towards cooperation and the conditions people place on cooperating with others.[28]

Expectations regarding cooperation are the foundation of morality in all human societies.[29] Such expectations may be codified as laws, but what really matters is how people actually behave — and how they feel everyone *ought to* behave. A law that is disrespected and flouted at every opportunity is a failure of legislation; to claim that breaking such a law is a sign of moral turpitude amounts to the application of an ethical standard that flouts the common-sense meaning of "ethics" and "morality" — words whose roots, respectively Greek and Latin, denote roughly "that, which people do."

At all times throughout the world tribes have supplanted other tribes; and as morality is one important element in their success, the standard of morality and the number of well-endowed men will thus everywhere tend to rise and increase.
— Charles Darwin (1871, p.159)

Real moral progress thus depends, first, on the possibility of developing from scratch, or identifying in the wild, a strain of ethics that is unquestionably better than the run of the mill; and, second, persuading enough people to adopt it, so that it *becomes* the run of the mill. Until such time, the best we can do is preserve the status quo, so that cooperative behaviors are not driven to extinction by those individuals who value personal gain over common good.

Given that cooperation places those who practice it at a fitness disadvantage relative to the "defectors" who don't, it indeed needs to be actively protected and preserved to prevent it from going extinct. This may happen on the level of biology, if the traits that enable it disappear from the population, or of culture, if cooperative behaviors become infrequent and can no longer be learned socially.

To redress this situation, something must be done to make it costly for defectors to free-ride, that is, to effortlessly benefit from the group's shared gain that stems from the cooperators' efforts. Typically, however, the punishment of defectors is itself costly for the punisher. This is why the success of this strategy, including "altruistic punishment" by third parties, in making cooperation generally worthwhile depends on the parameters that control the group dynamics. These include the costs of cooperation and of punishment, the benefit of cooperation to the group, the nature and frequency of inter-group conflicts (in which cooperating groups are more likely to prevail), and the patterns of migration (which may help spread prosocial behaviors across group boundaries).[30]

The complexity of group dynamics that takes into account all these effects makes it unwise to rely on intuition to predict how a particular setup would develop, or to do so by conventional mathematical means. Instead, evolutionary scientists and sociologists resort to running computer simulations of societies of interacting artificial agents. An agent-based model can simulate both biological and cultural evolution by endowing the agents with heritable traits as well as a capacity for social learning and transmission of behavioral patterns. Simulated behaviors may include foraging, sharing (a prosocial act), reciprocation, punishment, fighting, migration, and reproduction.

Agent-based modeling studies in computational evolutionary science and sociology — especially those that set the relevant parameters to values estimated from historical and anthropological data — have generally demonstrated that under a multi-level selection regime cooperation can emerge and remain stable over many generations, as long as the disruptive effects of individual defection are countered by some form of prosocial punishment.[31] This conclusion is in line with the characterization of human intuitive morality as conditional cooperation: we behave ethically on the condition that others do so too and that those who do not are penalized for it. But for these conditions to be verifiable, behavior — especially bad behavior — must be publicly observed. What, then, does one do if nobody is looking?

Conscience and Consciousness

If you ever pulled off in secret some mildly antisocial act and felt bad about it later, you know first-hand how morality works when nobody is looking. For the benefit of those whose antisocial escapades are habitually the opposite of mild, and who may therefore miss my meaning here, allow me to explain. The possibly quite punishing bad feeling is known as the "pangs of conscience," which may persist long after the offending event in the form of lingering guilt, shame, and regret.

Guilt and shame can, of course, also be sparked off by unconcealed misbehavior. What is impressive about these emotions and about conscience is that the typical human's capacity for self-punishment extends even to those transgressions, severe or slight, that are invisible to anyone but the culprit. The only way to account for this strange capacity is to accept that a conscientious person carries within them a special-purpose model of society at large, which incorporates its prevailing morals and can be effective enough at enforcing them.

Such a model[32] allows its host to tell how others would judge the host's act that has been, or is about to be, committed, whether or not it ends up being seen.[33] Our instinctive recognition of this future-oriented aspect of conscience

The English word "conscience" is derived from the Latin *conscientia*, originally defined as jointly knowing, knowing together with, or co-awareness, as well as consciousness and conscience. ... [This,] in turn, stems from the Greek term *syneidesis*, which refers to moral conscience, co-awareness of one's own bad actions, inner consciousness, accompanying consciousness, joint knowledge, or disconcerting inner consciousness.
— Frith and Metzinger (2016, p.211)

Contrary to the ideas of Dostoievski and the German apostles of inwardness, conscience consisted in the self's devotion to something substantial outside itself, in the ability to make the true concerns of others one's own.
— Horkheimer and Adorno (2002, p.164)

makes it a singularly useful tool in social decision making. In weighing the course of action in any situation that is morally charged, as any social behavior is, the anticipation of regret[34] can rule out a potential misdeed — or not, if one consciously chooses to defy the inner voice.

A thought-provoking account of the computational nature of the relationship between the self and the society, whose interests the self's inner voice represents, can be constructed by invoking the concepts of self-evidencing and the Markov blanket from chapter 3. Recall that for any cluster of physical processes to exist as a distinct entity, there has to be a Markov blanket, maintained through self-evidencing, that allows it to be individuated from everything else. Crucially, the mathematics that underwrites this account of existence is scale free.[35] As such, it applies not only to individual selves, but also to a set of individuals, if the latter is to have existence as community or group that is real, in the relevant sense, to itself and to other entities. This leads to an interesting perspective on sociocultural dynamics, according to which different groups, cultures, and countries are all trying to self-evidence. Insofar as this dynamics is driven by a dual imperative — to render the world unsurprising, by interacting with others who are similar to oneself, and, at the same time, expecting to be surprised so as to resolve uncertainty and learn about the greater world "out there" — it may help explain the tension between individualism and collectivism in a social species.

Members of a thoroughly domesticated and ultra-social species such as ourselves are best kept in line with tough love. Annoying as it may be when crossed, conscience fulfills a vital human need: integrating individual selves into the broader context of social consciousness. There are a couple of surprising aspects to this process, the first of which is that it works at all. While every society has its share of non-cooperators who cheat and free-load whenever they can, most people seem not to. Computational simulations of the social dynamics of conditional conscience-backed cooperation suggest that while the proportion of cooperating agents typically fluctuates over time, feedback loops can stabilize it at a high level for a broad range of the relevant parameters.[36]

The second surprising thing about conscience is that, paradoxically, the integration of self and society that it promotes is experienced consciously not as a reduction but as an expansion of personal freedom: in Ursula Le Guin's words, it liberates the individual from the "narrowness of the single self." This insight may help us understand the intention behind the words of Confucius and of Goethe that serve as epigraphs for this chapter.

And as understanding grew between us and we came to trust each other, there were no restraints on me at all but those of religion and my duty to my people. I had grown up with those, they were part of me, not external, not enslaving; rather, in enlarging the scope of my soul and mind, they liberated me from the narrowness of the single self.
— Ursula K. Le Guin (2008, p.185)

Significantly, conscience in all of its manifestations, including guilt and shame, is necessarily experienced consciously. It is also to be noted that the very possibility of conscience critically depends on the presence of a social context — that is, of agents-patients other than oneself — without which there can be no meaning to the concept of morality. Because conscience does exist and is what it is, the focus of this chapter — social consciousness — must be recognized as "the interface between the person and culture."[37]

The Fifth Revolution

For a deeply social species, whose members must continually choose between cooperation and hardship, deprivation, or death, any development that promotes cooperation should by rights be a welcome one. The fifth consciousness revolution stabilized and enhanced cooperation by bringing us a new kind of self-awareness, into which our complex relationships with others are integrated. It has also exposed us to new kinds of suffering, among them the pangs of conscience and the pain of social exclusion. Arguably, though, this latest vulnerability is more than made up for: first, by the intuitive confidence that we have acquired in the natural goodness of most of the people out there, and, second, by people being in fact mostly good.

rén zhī chū
xìng běn shàn
(people at birth
are naturally good)
— *Sānzi Jīng*
(*The Three-Character Classic*,
Song Dynasty)

That people are naturally good at birth counts for nothing if their goodness is not tended to by their social milieu. The opening lines of *The Three-Character Classic*, quoted in the margin, are immediately followed there by a disclaimer: while people's natures are similar, their habits make them different from each other.[38] It makes a hell of a lot of difference where these habits come from. Because people are *conditional* cooperators, raising children in a harsh environment, where competition is rampant and mutual aid scarce, would undo any innate predisposition to goodness that they may have. Conscience, powerful as it can be, needs a moral example to imprint on; and any such examples can only come from without. One can only hope — for everyone's sake, as we are all in this together — that everyone gets to learn from good role models and not from bad ones.

The Counter-Revolution, or: What Went Wrong

It is interesting how in this book, once we got to human-like consciousness, the concept of *hope* keeps cropping up in the discussion. At the end of chapter 4, I have called language a vehicle of hope; now, nearing the conclusion of chapter 5, we can and must start thinking about filling it with content. What hope *is* there for coexistence and cooperation?

There is no doubt that
something has gone terribly
wrong with the world. A very
small percentage of its
population do control the fates
of almost everyone else, and
they are doing it in an
increasingly disastrous fashion.
— Graeber and Wengrow
(2021, p.76)

Right now, things look rather bleak. As my hardscrabble, pogrom-dodging ancestors, who knew a thing or two about hoping against hope, might have muttered at this point — in Yiddish, a language perfectly suited both for mockery of humbug and for complaining about life — *yener tsuzamenlebnkeit hobn mir do*: some coexistence we have here.

Indeed, our capacity for cooperation seems to be on the verge of failing major tests even on issues that one would imagine to be of universal concern and likely consensus, such as basic human rights and global warming. A handful of corporate oligarchs, fascist demagogues, "people's party" preceptors, and warlords preside, virtually unopposed, over a world order that is literally the stuff of nightmares. Worse, this world order appears to be eerily stable, despite some or other of its empires and nation states being at all times in the middle of an economic crisis, or a war, or both.

All this is very strange, given the massive amounts of evolutionary, archaeological, historical, and anthropological evidence for cooperation being our species' defining trait and birthright. The stark contrast between our collective potential for goodness and our sorry state begs a question, which Graeber and Wengrow keep returning to in *The Dawn of Everything*: What Went Wrong?

Clearly, the achievements of the slow revolution of consciousness that had made us good at birth and for the most part kept tyranny at bay came to be opposed by powerful political-economic forces. While a proper analysis of this regrettable development (arguably the most consequential one in human politics to date) is outside the scope of this book, a rough account of what went wrong can be pieced together and may be worth pondering, if we are to make sense of our present plight.

In traditional societies, attempts on part of an individual to exercise undue power over others were typically met with collectively orchestrated and socially sanctioned resistance, aimed to preserve a *reverse dominance hierarchy*.[39] The means of resistance, as documented by anthropologists, varied widely, ranging from conspicuous disobedience and public ridicule to banishment or killing of the offender. With time, the valor of tyrant-killers came to mean less and less. Two hundred years ago, Alexandre Pushkin could still inspire his readers by praising Brutus's dagger:

A key event in early human evolution was a form of guarded egalitarianism that made it difficult for some individuals to dominate others within their own group.
— D. S. Wilson & E. O. Wilson (2007, p.343)

> Свободы тайный страж, карающий кинжал,
> Последний судия позора и обиды.
>
> (The secret guardian of liberty, the punishing dagger
> The ultimate arbiter of shame and insult.)

Yet, the killing of Julius Caesar by the "freedom-loving" Brutus proved to be merely a desperate paroxysm of the Roman republic as it was being remade into an autocracy. Five years after Pushkin wrote *The Dagger*, the 1825 Decemberist parade-ground "revolt" failed to convince the young czar Nicholas of the need to liberalize (Nicholas I went down in history as Николай Палкин, "Nicholas Canin," because of the widespread use of judicial caning during his long reign). The Romanovs would not face a more persuasive argument against autocracy until 1881, when Alexander II was blown up by members of the revolutionary movement called *Narodnaya volya* ("people's will" or "people's freedom") — only to be succeeded by Alexander III.

Already in Caesar's Rome, then, it was too late to fight tyranny with daggers. The reason for this has to do with the very nature of the centralized state. Whether they are dictatorships or democracies, states are characterized by a deep disconnect between the populace and the apparatus of political control. That this apparatus is itself typically controlled by a tiny minority muddles the people's class consciousness and misdirects their attention and ire away from the real cause of the condition, which is systemic and structural.

The resilience of the statist social structure stems from the intangibility of the machinery of state itself — that, which remains when the particulars of the ruling elite (strongmen, captains of the industry, oligarchs, party cadres, etc.) are abstracted away. Just like in evolution in general, in political-cultural evolution too nothing succeeds like success: with the machine of state up and running, there is no place where sticking a dagger can do much good, no matter how gratifying it would always be to keep a despot quaking in his caligae by a constant threat of retribution.[40]

The powers that guard this machine need not themselves be invisible or intangible — on the contrary. On my first visit to Seoul, in 2008, I was shown around by one of my host's graduate students. Along a tree-lined avenue leading to the government complex, we saw a dozen parked police buses, each full of terminally bored-looking teenagers in ill-fitting uniforms. I asked my guide whether in Seoul too the government uses very young and under-educated recruits from the periphery to police the capital. The student politely ignored my question, but a Korean friend later confirmed my guess.[41]

The Korean approach to policing the all-important state capital illustrates the three factors that together make modern states (as opposed to political regimes, which come and go) so resilient. The first one is the development of a taste for power on part of the *ruling class*, which then eagerly uses its resources to prop up any useful autocrat. Given the abundance of sources that offer a class analysis of world history,[42] there is no need to belabor that point here. The other two factors are closely related to each other. They are the adoption of *mercenary enforcement*, which exempts the ruler from having to rely for security on the same people whom he aims to subjugate; and the invention of *money*, which the ruling class can exact in tribute and use to pay its mercenaries.

Mercenaries can be inexpensive if they are drafted instead of being paid. The case that prompted my question about using draftees for policing Seoul was that of the Israeli Border Guard force, which is composed of Jews and minorities such as the Druse, all drafted as teenagers and deployed to the occupied territories and Arab-majority areas. Examples of more traditional mercenary forces that come to mind are the British colonial military and paramilitary forces, such as the Gurkhas, the Irish Royal Constabulary, and the Black and Tans. Closer to home, there is the U.S. police, who tend to live in suburbs, while being paid, extravagantly, to preserve "law and order" in inner cities.[43]

A striking historical instance of imported mercenary power is the tale of Rurik the Varangian, the semi-legendary founder of the dynasty that ruled Russia through the end of the 16th century. According to the story, several Slavic tribes, whose land was "rich, but lacking in order," invited Rurik to come and rule over them, which he promptly did, accompanied by his own army. The historical accuracy of this story is less important than the Russian chronicler's choice to preserve it.[44] A notable true story from Russia's more recent history is the czars' use of Ukrainian and Don Cossacks in its imperial push into Siberia, as well as south into the steppes, northern Caucasus, and Chechnya, the latter ironically mirrored by the recent recruitment of turncoat Chechens as Russia's mercenaries in its aggression against Ukraine.[45]

The practice of relying on non-local mercenaries, or settlers of some kind or another, in subduing and keeping in line a hostile local population seems to be universal. It does make a lot of sense for a would-be autarch to import foreign troops to defend his rule — or, even better, to *be* a foreigner who arrives with his own troops and owes no allegiance to the people he rules over. It's hard to reverse a dominance hierarchy that is guarded by strangers whose salary is paid by the top dog and who may not even speak your language.[46]

[Money] tends to be identified with the holder's generic, *hidden* capacities for action.
— David Graeber (1996, p.6, my emphasis)

Mercenaries and money draw their special powers from the same phenomenon: selective *invisibility* (which also contributes to the intangibility of the machinery of state, mentioned earlier). On the one hand, imported mercenaries are more effective locally than an indigenous force insofar as they do not see the locals as real people and are therefore easier to have their conscience silenced by money. On the other hand, for a tyrant whose face is only ever seen on the coins he mints, money is easy to wield in recruiting military support, especially if the recruits come from elsewhere and have no local family or friends to act as their external conscience. In a 1844 manuscript titled *The Power of Money*, Marx wrote that "he who can buy bravery is brave, though he be a coward," to which one might add that he who buys braves abroad can be brave at home with more than usual degree of impunity.[47]

People have long been fascinated by stories about the potential for impunity afforded by invisibility, when the only remaining barrier against immoral (that is, antisocial) behavior is the person's conscience. In Plato's *Republic*, the debate about morality is anchored by the story of Gyges, whose possession of a magical ring that made its wearer invisible helped him usurp the throne of Lydia.

The invisibility of absolute power and the anonymity of money are intimately intertwined in this story, a version of which is found also in Herodotus. Marc Shell, who discusses both versions in *The Ring of Gyges*, writes:[48]

> As coinage was associated with Lydia, so too was political tyranny, "a phenomenon no less important in the history of culture than in the development of the Greek state." The very word *tyrannos* is Lydian in origin. Many Greeks believed that Gyges was the first tyrant and often associated him with tyranny; he was the archetypal tyrant as he was the archetypal minter. Indeed, the frequent association of tyranny and minting with one man suggests that they may be mutually reinforcing and interdependent.

It was the invisible hand of money in politics that the Greeks had feared as a prime instrument of tyranny. With the benefit of hindsight, we can conclude that any ideology that holds an invisible hand to be a good thing should at the very least be approached with suspicion.

An interesting lesson in the politics of invisibility has been taught to the American public by the U.S. Supreme Court in its 2010 decision in the *Citizens United v. Federal Election Commission* case. By allowing anonymous entities to spend unlimited money on swaying elections, the court removed the last restraint that had kept capital from buying political power in this country. At the same time, the ruling made many people conscious of the incongruity in calling a system that lets *any* money into politics a "democracy." The more adventurous among corporate pundits then started using the phrase "dark money" without acknowledging the redundancy: as a long line of thinking in political philosophy, stretching from Plato to Marx and beyond, suggests, all money is dark money.[49]

All MONEY is dark money.

One of the key functions of the modern state is ensuring that its citizens are sufficiently motivated to pursue and spend money with appropriate enthusiasm. A community that grows or manufactures everything that it needs denies profits to those in power; as such, it is a danger to the state — if word gets out, to states everywhere — and must be destroyed, or at least reeducated.[50] The same goes for people who have already been assimilated into the money machine, but fail to spend enough, either because they don't need more than they have, or because they don't earn enough in the first place. Such people must be persuaded that they need more, or, as the case may be, compelled to work harder, so that they can afford to buy more.

In an age when the use of arable land or pasture was the most important prerequisite for independent subsistence, the military might of the state helped the gentry amass estates (and capital) by taking over smallholders' plots and especially the land held in common, by a process called *enclosure*.[51] These brutal but of course perfectly legal acts have been particularly closely studied in England, where they transformed land ownership and have been the main cause of peasant revolts for centuries. Enclosures vastly enriched the landowners, dispossessed peasants and small tenants, and created a contingent of utterly poor people who were forced to sell their labor and who would eventually become the urban working class.

England's colonial project operated on the same principles, on a much larger scale. While the original push to acquire overseas possessions had been motivated mostly by the natural resources that could be extracted and brought back home or sold to third parties, the imperial administration eventually realized that its subjects in the colonies could also be made to generate income by having them owe taxes. Because self-sufficient agrarian communities, such as those in early colonial India, did not have any disposable income, levying a tax on them forced people to abandon their traditional cultivars in favor of cash crops such as tea or sugar cane. This also made it easier to press the locals into buying British cloth and industrial products, for which there would otherwise be no *market*.[52]

The concept of market is the last brick that was missing from the foundation for the present informed guess, such as it is, regarding What Went Wrong in recent human history:

- First, as Graeber and Wengrow have shown in *The Dawn of Everything*, at all times and in every part of the world, many different types of social organization could be found, some of which happened to be, for a variety of reasons, more centralized that others.
- Second, synergy among several factors — notably, nascent upper-class interests, the invention of money as we know it, and the possibility of recruiting a force answerable to the state rather than to its citizens — then promoted the emergence of local powers.
- Third, the ruling class found ways to rapidly increase its power through enclosures, dispossession of most of its own people at home, and, if at all possible, imperial expansion abroad.
- Fourth, the institution of markets for goods and for money paved the way for the political-economic system that presently dominates the world stage: *capitalism*.
- Fifth, the people's individual and collective consciousness has been subverted and immured in a virtual consensus, which holds self-interest, competitiveness, aggression, and greed to be both true descriptors of human nature and normatively preferred values for managing human affairs.

This emergence and entrenchment of false consciousness among the planet's population explains the ability of capitalism — protected and promoted by state power — to withstand periodic assaults in the form of civic unrest, wars, revolutions, and the inevitable economic crises of its own making (the latter having been predicted already by Marx in *The Capital*).

In the 21st century, capitalism appears to be nearing its apotheosis. In the West, this has been heralded by the U.S. Supreme Court's affirmation of corporate personhood (in the *Citizens United* ruling, which I mentioned earlier). The newly empowered American and transnational "corporate citizens" are, by any traditional community standard, not only immoral,[53] but, arguably, psychopathic.[54] The situation is, if anything, even more dire in "communist"

Esprit de corps, national sentiment, sympathy, are no mere metaphors. None of us can fully realize what the minds of corporations are, any more than one of my brain cells can know what the whole brain is thinking. But the law of mind clearly points to the existence of such personalities.
— Charles Sanders Peirce (1892)[55]

Twenty-first century capitalism as a whole is a zombie system, seemingly dead when it comes to achieving human goals and responding to human feelings, but capable of sudden spurts of activity that cause chaos all around.
— Chris Harman (2010, p.12)

China, where the ultimate corporation that is the totalitarian state tolerates predatory capitalist self-enrichment by some of its citizens (as long as they submit to the political control of the party bosses), while keeping hundreds of millions of peasants and workers in near-poverty.[56]

These characteristics of the corporate mind make it unlikely to be receptive to moral persuasion. Moreover, even if corporations are conscious (as suggested by some philosophers),[57] their consciousness necessarily and radically differs from ours, insofar as it is not embedded into a social matrix that shapes and sustains human consciousness. In practical terms, therefore, corporations should be treated just like philosophical zombies — a conclusion that resonates with Marx's view of capital as "dead labor, which, vampire-like, lives only by sucking living labor, and lives the more, the more labor it sucks," as well as with the more recent notion of zombie capitalism.[58]

A world infested with zombie corporations, whose hunger for natural resources, labor, and sheer power can never be sated, is not a good place to be. Harry Brighouse, in his comments on Erik Olin Wright's paper *Transforming Capitalism Through Real Utopias*, offers a list of "*specific ways* in which capitalism is *bad*" (his italics):[59]

- Capitalist class relations perpetuate eliminable forms of human suffering.
- Capitalism blocks the universalization of conditions for expansive human flourishing.
- Capitalism perpetuates eliminable deficits in individual freedom and autonomy.
- Capitalism violates liberal egalitarian principles of social justice.
- Capitalism is inefficient in certain crucial respects.
- Capitalism has a systematic bias towards consumerism.
- Capitalism is environmentally destructive.
- Capitalist commodification threatens important broadly [and, presumably, rightly] held values.
- Capitalism, in a world of nation states, fuels militarism and imperialism.
- Capitalism corrodes community.
- Capitalism limits democracy.

It is worth noting that seven out of these 11 characteristics of capitalism are directly predicated on consciousness.[60] Of these, the most consequential one (and the first one listed) is *eliminable suffering*. At long last, we are now in a position to give this aspect of the human condition the attention it deserves.

❧ ❧ ❧

Notes

1. The Confucius quote is from *Analects* (II, 4). For some historical and philosophical background to the Confucian views on the place of an individual in society, see *A Short History of Chinese Philosophy* by Yu-Lan Fung (1966). Li (2014) discusses specifically the Confucian concept of freedom, which the quoted passage illustrates so vividly.

I came across the Goethe quote (from *Bekenntnisse einer schönen Seele*), in the original German, in the first of Hakugen Ichikawa's two papers on Zen ethics (Ichikawa, 1963, 1965), where he compares Schiller's concept of *schöne Seele*, (beautiful soul) to the Buddhist practice of *prajñā* (selfless intuitive intelligence) and Zen's "preceptless precepts." This quote from Goethe, which Ichikawa (1963, p.356) chose to illustrate the European Romantics' conception of being good by natural inclination (as opposed to merely obeying societal rules) reads strikingly like a paraphrase of Confucius. Lastly, the "What Went Wrong?" quote is from (Graeber and Wengrow, 2021, p.1).

2. The "boyish plan" quote is the last stanza (less the very last line) from the poem *What Then* by Yeats (1966, p.184). Elangovan and Hoffman (2019), who also quote Yeats, remark (p.71) in this connection on the relationship between the individual (academic) and society:

> Our interpretation of academic scholars as stewards of societal well-being is a point of departure from March's (2003) protestation that "a university . . . is a place where learning and scholarship are revered, *not primarily for what they contribute to personal or social wellbeing* but for the vision of humanity that they symbolize, sustain, and pass on" (p.356; italics added for emphasis). We draw our inspiration from John Donne's (1624) thoughtful essay *Devotions upon Emergent Occasions* in which he noted, "No man is an island, entire of itself" (*Meditation XVII*).

3. For my take on regret, see the eponymous chapter 28 in (Edelman, 2020). To forestall the obvious objection: yes, I am aware that any time that regret is voiced by a person as successful as I am (by the usual American standards), it is always from a position of extraordinary privilege.

4. I write "presumed" because, of course, human political-economic imagination does not conform to the Cold War conception of a unidimensional communism-capitalism axis (nor, for that matter, to the liberal-conservative distinction so beloved of the Western corporate media). It took me several decades to understand this.

5. See (Edelman, 2008, section 9.7.1), *Solitude*.

6. The rather unsurprising bottom line of that book (Edelman, 2012) was that the process of pursuing happiness matters more than the fleeting moment of its attainment.

7. In the U.S. economy, the promise of upward mobility in return for hard work has been aptly characterized by Chetty et al. (2017) as "the fading American

dream." As to the propaganda value of "work ethic," Graeber (2020, p.166) remarks that it has taken on religious overtones: "Work is a kind of secular hair shirt." I will return to discuss the conscious experience of a wage worker in chapter 7, in the context of work-induced alienation (Sawyer and Gampa, 2020).

8. Paulo Freire, himself a great educator, had no illusions about the power of education on its own:

> [E]ducation is not a miraculous process capable by itself of effecting the changes necessary to move a nation from one epoch to another. Indeed, it is true that by itself education can do nothing, because the very fact of being "by itself" (i.e., superimposed on its context) nullifies its undeniable power as an instrument of change.

The quote is from (Freire, 1974, p.36).

9. I was also unlucky in the selection of books in my local second-hand bookstore. I like to think that, had I come across any of Ursula Le Guin's books in my twenties or even thirties, my miseducation would have been more than compensated for in time for me to choose a more socially engaged career than what is typically afforded by degrees in electrical engineering and computer science.

10. The reading list for my *Varieties of Freedom* seminar (available on request) included perspectives from China (Li, 2014; Wu, 2006), Brazil (Lino e Silva and Wardle, 2017), Nigeria (Kanu, 2017; Oyeshile, 2007), Britain (Berlin, 1969), and India (Chakrabarti, 2018), as well as cross-cultural anthropological reviews such as (Lee, 1963).

11. Concerning the (in)viability of isolated individuals, Dobzhansky (1964, p.444) writes: "Individuals and species belong to ecological communities and ecosystems. An individual taken out of the system in which it normally occurs is incomplete and it may be inviable."

12. For information on the geography and history of Death Valley, see (Digonnet, 2016).

13. This quote, possibly from notes prepared for a public lecture, can be found in (Arendt, 2017, pp.63-64). Elsewhere, she writes that "We first become aware of freedom or its opposite in our intercourse with others, not in the intercourse with ourselves. ... in order to be free, man must have liberated himself from the necessities of life" ("What is Freedom?", Arendt, 1961b, p.149). The economic preconditions for freedom have been discussed at length by Murray Bookchin (1986a; 1982; 1986b) and, more recently, by David Graeber (2020).

14. Concerning the jungle metaphor, see Valverde (1996, pp.503-504): "The 'jungle' is not a single signifier with a single meaning: it is a very rich allegory containing quite contradictory desires and meanings."

15. In political philosophy, the concept of state of nature is central to the work of Mozi (470-391 BCE) in the Warring States China, and of Thomas Hobbes (1588-1679) in England. The line "Nature, red in tooth and claw" is from Alfred Tennyson's 1850 poem *In Memoriam A.H.H.*

16. Concerning the history and the anthropology of states, see for instance (Clastres, 1987; Graeber, 2004; Scott, 1998), and of course *The Dawn of Everything* by Graeber and Wengrow (2021), to be discussed later in this chapter.

17. The quote is from (Ward, 2004), available online (hence no page number).

18. I found this quote from Élisée Reclus (*L'Homme et la Terre*, Vol.I, pp.140-141. Paris: Librairie universelle, 1905) in Ferretti (2017). Also relevant here is Herbert Spencer's view of evolution as immanent progress: "the intersection of Spencer's evolutionary anarchism and classical liberalism: social Darwinism bridges the divide" (Colson, 2011, p.37).

19. Kropotkin's *Mutual Aid* (1902) is available online at https://theanarchistlibrary.org. Servigne and Chapelle (2022) have recently brought Kropotkin's project up to date with a wealth of evidence for what he had termed mutual aid, from many species and ecosystems.

20. My conception of cooperation follows Boyd and Richerson (2009, p.3283): "we use the word cooperation to mean costly behaviour performed by one individual that increases the payoff of others"; see also (Henrich and Muthukrishna, 2021) on "the origins and psychology of human cooperation." An overview of the Extended Evolutionary Synthesis can be found in (Laland et al., 2015).

21. This quote is from (Kropotkin, 1902, Chapter 3).

22. A preview of the monumental *Dawn* can be found in Graeber and Wengrow's earlier collaboration, *How to change the course of human history (at least, the part that's already happened)* (Graeber and Wengrow, 2018). The consciousness quote is from (Graeber and Wengrow, 2021, pp.93-94).

23. Burns and Engdahl (1998, p.166) argue that "individual consciousness results from the development of collective naming, classifying, judging, reflection, and discourse applied to individual members or participants in the collective." Clearly, this take is not applicable to any consciousness other than human, and as such seems overly narrow. For humans, however, the claim that "personal identity is social identity" (the title of Carr, 2021) does make sense — as the line of reasoning developed in the present chapter suggests.

24. The titles of these two papers say it: "Wolves, but not dogs, are prosocial in a touch screen task" Dale et al. (2019); "Wolves outperform dogs in a conspecific cooperation task" (Marshall-Pescini et al., 2017).

25. The full saying (in the original Latin, *Lupus est homo homini, non homo, quom qualis sit non novit*) is attributed to T. Maccius Plautus and dates back to republican Rome.

26. Boyd and Richerson (2009) quote several passages on the evolution of cooperation from *The Descent of Man* (Darwin, 1871): one that points out the expected downside of altruism on the personal level ("he who was ready to sacrifice his life, as many a savage has been, rather than betray his comrades, would often leave no offspring to inherit his noble nature"; p.163), and a complementary one, which posits that altruism may be beneficial on the group level ("a tribe including many members who ... were always ready to give aid to each other and sacrifice themselves for the common good, would be victorious over most other tribes; and this would be natural selection"; p.166).

27. The quote is from (Wilson and Wilson, 2007, p.343). Concerning the evolutionary origins of politics, see (Gintis et al., 2015).

28. One of Marx's peculiar phrases for human nature was "species being," for which Dean (2016, p.509) offers the following gloss: "Marx equated it [species being] with the materiality of history. In doing so he was alluding not merely to the objective materiality of nature, but the materiality of human, and therefore social, action; of 'revolutionary, or practical-critical activity' (Marx, 1845: 82)". Morgan (2018) revisits "species being in the 21st century."

29. Massive evidence for the universality of cooperation-based morality across human societies is available (Alfano et al., 2022; Curry et al., 2019).

30. These characteristics of selection by group apply also to "levels" of population (Su et al., 2022).

31. See (Odouard et al., 2021) and the references therein. Note that withholding of cooperation is a type of punishment.

32. The idea that individuals incorporate models of the society in which they are embedded has been put forward, for instance, by Frith and Metzinger (2016, p.212): "Here, we propose that the "ideal observer" — which lies at the origin of moral cognition and moral behavior — is a mental representation of group interests. This is the emergence of a "first-person plural perspective" (Gallotti and Frith 2013). Self-consciousness served as a functional platform for the representation of group preferences in the brain of individual organisms. Upon this platform, individual and group interests could compete. The mechanisms which constitute self-consciousness are subpersonal; the representational content is often suprapersonal."

33. The idiosyncrasy of of morality is illustrated by the (unsurprising) finding that "children across societies enforce conventional norms but in culturally variable ways" (Kanngiesser et al., 2022).

34. For computational theories of counterfactual regret minimization, see (Brown et al., 2019; Hoerl and McCormack, 2016); the social psychology angle is represented by (Breugelmans et al., 2014; Davidai and Gilovich, 2018; Zeelenberg et al., 2002). Frith and Metzinger (2016, p.208), who tie consciousness to conscience, write that "... regret always has to do with conflicting sets of preferences and its representational content is inherently social. In essence,

regret results from applying mechanisms of social control to oneself, namely, retribution (self-punishment) and reputation loss (self-blame)."

35. I am grateful to K. Friston for pointing out to me the scale freedom of the mathematics of Markov blankets (Fields et al., 2022; Friston, 2019) and its relevance to the ontology of societies of selves.

36. House et al. (2020) discuss the contribution of third-party punishment to enabling cooperation. Agent-based simulations of the effects of conscience (construed as internalized morals) on the dynamics of cooperation have been carried out by Odouard et al. (2021). Among the specific hypotheses concerning the evolution of conscience, a particularly intriguing one ties it to parent-offspring conflict (Voland, 2014). On the level of mechanism, conscience, being socially induced pain, shares brain circuits with other types of pain (Walters and de C. Williams, 2019).

37. In their paper titled *What's the use of consciousness? How the stab of conscience made us really conscious*, Frith and Metzinger (2016, p.212) refer to "consciousness as the interface between the person and culture" and link it to the concept of virtual self (cf. Metzinger's (2003, p.553) notion of "total flight simulator," which I discussed on p.67):

> The term "virtual identity formation" was introduced to refer to this process (Metzinger 2013:5) and it is speculated that one function of mind wandering is the constant creation and functional maintenance of the representation of a transtemporal, fictional "self." Only if an organism simulates itself as being one and the same across time will it be able to represent reward events or the achievement of goals as happening to the same entity, as a fulfillment of its own goals.

Regarding the claim that there can be no morality without a social context: a more precise formulation would be "without other moral agents than oneself"; arguably, the category of moral agent includes all entities capable of suffering, thus also non-human animals, and perhaps all "sentient beings" (to use the Buddhist expression).

38. Fung (1966) quotes from *The Three-Character Classic* on p.1 of his volume.

39. The concept of reverse dominance hierarchy was introduced by Boehm et al. (1993).

40. The use of personal terror for political purposes has a rich history in Russia. Vladimir Ulyanov-Lenin's elder brother, Aleksandr Ulyanov, a member of *Narodnaya volya*, was hanged in 1887 for his participation in a plot to assassinate the reactionary Tsar Alexander III of Russia (on the 10th anniversary of the blowing up of Alexander II). Following his brother's execution, the young Vladimir is reported to have renounced personal terror as ineffectual: "Нет, мы пойдём не таким путём. Не таким путём надо идти." — No, we shall not go down that road. That's not the road one should choose (as recounted by Lenin's elder sister Anna in a memoir published in 1962). Thirty

years later, Lenin's Bolshevik faction of the Russian Social Democratic Labor Party seized power in the October 1917 revolution. Immediately, it commenced a bloody terror campaign, not only against the supporters of the old regime, but also against communists and anarchists who opposed the absolute dictatorship of Bolshevik apparatchiks, propagandized as the "dictatorship of proletariat." This made Lenin a target of assassination attempts, of which the one by Fanny Kaplan of the Socialist-Revolutionary Party, in 1918, came the closest to succeeding. One of the architects of the Bolshevik terror, Leon Trotsky, was himself assassinated on Stalin's orders in 1940.

41. I thank Juanita Hong for expert advice in this matter.

42. A couple of useful starting points for this inquiry are (Gran, 2007) and (Robinson and Barrera, 2012).

43. Karpiak and Garriott (2018, p.1) mention briefly the historical origins of the modern police in the UK. For a review of the anthropology of security, see (Goldstein, 2010).

44. While the historical veracity of the classical tale of Rurik and the "Invitation to Varangians" (Riasanovsky, 1947) has been rightly questioned (Harris and Ryan, 2004; Spears, 2018), what matters is the persistence of this story in the popular imagination of the people involved. A similar sentiment is voiced by Shell (1989, p.22) in connection to the other famous tale that I invoke here: "The story of Gyges, however hypothetical or mythical, is a great explanation of the genesis of a political, economic, and verbal semiology." A historically reliable account of the role of money and mercenaries in Mesoamerica can be found in (Fargher et al., 2011).

45. For the historical context of the Russian-Chechen conflict and in particular the recent role of Kadyrov's mercenaries, see (Bielicki, 2018; Souleimanov et al., 2019). Wither (2020) discusses the outsourcing of warfare.

46. Rulers' habit of using foreign troops to keep the populace under control has been noted by David Graeber in the essay *There never was a West, or, Democracy emerges from the spaces in between* (2007):

> [T]he term "democracy" ... appears to have been coined as something of a slur by its elitist opponents: it literally means the "force" or even "violence" of the people. Kratos, not archos. The elitists who coined the term always considered democracy not too far from simple rioting or mob rule; though, of course, their solution was the permanent conquest of the people by someone else.

47. Here's an invective directed at money by the young Marx (1961, XLIII):

> Money, then, appears as this *distorting* power both against the individual and against the bonds of society, etc., which claim to be *entities* in themselves. It transforms fidelity into infidelity, love into hate, hate into love, virtue into vice, vice into virtue, servant

> into master, master into servant, idiocy into intelligence, and intelligence into idiocy.

A connection between money and false consciousness, another concept closely studied by Marx and Engels (Augoustinos, 1999), has been made already by Plato, who "studied the 'false consciousness' in which money appears to be that which it is not" (Shell, 1989, p.44).

48. According to Shell (1989, p.22), invisibility is a key ingredient of the Greek conception of a tyrant, who is not seen, but who sees everything.

49. On this recent turn in political commentary, Harman (2010, p.351) remarks: "As I write this, the sheer scale of the crisis facing the world system is forcing even those who run the system to talk about capitalism, and having to recognise that Marx had something to say about it long before Keynes."

50. A fascinating fictional treatment of the state's backlash against those who would leave it behind and try to build a new, better world is the novel *Walkaway* by Cory Doctorow (2017).

51. Enclosures (Chatterton and Pusey, 2020; Fairlie, 2009; Hickel, 2019) were decried by Thomas More in *Utopia* (1516). An anonymous 18th century poem sums up the popular view of their effect as follows:

> The law locks up the man or woman
> Who steals the goose from off the common,
> But lets the greater felon loose
> Who steals the common from the goose.

Fairlie's *Short History of Enclosure in Britain* (2009) shows how different the process of state-orchestrated enclosures was from the mythical "tragedy of the commons," invented and promoted by neoliberal ideologues. In his book *Zombie Capitalism: Global Crisis and the relevance of Marx*, Harman (2010, p.39) refers to the stealing of the commons as the "primeval act of robbery — the use of force to remove masses of people from any control over the means of production. This was often carried through by the state at the behest of some of the most privileged groups in society. In England and Wales, for example, the rise of capitalism was accompanied by "enclosures" — the forcible driving of peasants from common land they had cultivated for centuries."

52. See for instance the analysis of the ecological and economic effects of tea as a colonially enforced cash crop in India (Sarkar, 2018).

53. Dilevko (2014) offers some examples of the "demise of morality" under the corporate state.

54. Here is how Mulgan (2019, p.906) describes the psychopathy of corporations:

> A familiar theme in the recent literature is that many actual corporate groups — and especially for-profit business enterprises — are closer to psychopaths than to normally functioning human adults

> (e.g. Bakan 2004, p.56ff). They can act, intend, plan, and display instrumental rationality. But they feel no empathy, cannot recognise the moral importance of others' points of view, and cannot recognise (and therefore cannot respond appropriately to) moral reasons. Indeed, psychopathic behaviour is encouraged by modern theories of the firm that insist business enterprises should single-mindedly maximise returns to shareholders! (Nor is legal incorporation any barrier to psychopathy, because modern corporations can pursue any legal purpose.)

In this connection, it is interesting to consider the conclusions from an empirical investigation of "the opposite of psychopathy," which Woodmass and O'Connor (2018, p.254) found to be defined by

> ... a tendency to experience guilt following private transgressions, by a tendency to experience shame following publicly exposed transgressions, by honesty, humility, cooperativeness, agreeableness, consideration, restraint, and conscientiousness. ... "Compassionate morality" is suggested as a possible label.

55. Lane (2009, p.13) quotes Peirce's arguments in favor of the possibility of corporate consciousness, while disagreeing, for the large part, with his conclusions.

56. An analysis of China's transition from socialism to capitalism can be found in (Harman, 2010, p.241, chapter "Hope in the East?"). A revealing aspect of that transition was the phasing out of the guaranteed subsistence income policy, nicknamed the "iron rice bowl" (Harman, 2010, p.243; see also Easterlin et al., 2012, p.9777).

57. The question of proper criteria for consciousness in "hive minds" is problematic, as discussed, for instance, in (Fekete et al., 2016).

58. The "vampire" quote (Marx, 1887, p.163) is one of several such expressions in *The Capital.* The phrase "zombie capitalism" is part of the title of (Harman, 2010).

59. The Brighouse (2021, p.5) quote on capitalism is from his paper on (Wright, 2013). Ci (1999) discusses cultural aspects of capitalism. Sinclair (1976) and Robinson and Barrera (2012) expose the close links between capitalism and fascism. The roots of capitalism in slavery are noted by Graeber (2004, p.71): "[...] modern capitalism is really just a newer version of slavery. Instead of people selling us or renting us out we rent out ourselves." In this connection, Graeber (2006, p.66) takes care to note that "The idea that capitalism is as old as civilization is of course a position long popular amongst capitalists[.] [...] The problem of course is that, defined so broadly, it becomes very hard to imagine eliminating capitalism at all."

60. I close this chapter's overview of rarely considered aspects of consciousness — in particular, consciousness under capitalism — with a quote from Horkheimer and Adorno (2002, p.175), who write:

> If the materialist critique of society once opposed idealism by asserting that it is not consciousness which determines being but being consciousness, and that the truth about society is to be found not in its idealistic notions of itself but in its economy, up-to-date self-consciousness has meanwhile discarded such idealism. People judge their own selves by their market value and find out who they are from how they fare in the capitalist economy.

References

Alfano, M., Cheong, M., & Curry, O. S. (2022). Moral Universals: A machine-reading analysis of cultures. *OSF*. https://doi.org/10.17605/osf.io/86qry

Arendt, H. (1961a). Freedom and politics. In A. Hunold (Ed.), *Freedom and serfdom: An anthology of Western thought* (pp. 191–217). Dordrecht, Holland: D. Reidel.

Arendt, H. (1961b). What is freedom? In *Between past and future: Six exercises in political thought*, Chapter 4 (pp. 143–172). New York, NY: Viking Press.

Arendt, H. (2017). The freedom to be free. *New England Review*, *38*(2), 56–69. A lecture delivered in 1966-7.

Aristotle. (1885). *Politics*. Available online at http://classics.mit.edu/Aristotle/politics.html.

Augoustinos, M. (1999). Ideology, false consciousness and psychology. *Theory & Psychology*, *9*(3), 295–313.

Bakunin, M. (1871). About freedom. In A. Lehning (Ed.), *L'Empire knouto-germanique et la Révolution sociale, 1870-1871*. Leiden: Brill.

Berlin, I. (1969). Two concepts of liberty. In *Four essays on liberty* (pp. 121–154, 169–172). Oxford, UK: Clarendon Press.

Bielicki, P. (2018). The significance of Russian-Chechen relations for contemporary international security. *Nowa Polityka Wschodnia*, *4*, 117–140. In Polish.

Bloch, E. (1971). *On Karl Marx*. New York, NY: Herder and Herder. Original edition: *Über Karl Marx*, Suhrkamp Verlag (1968), from *Das Prinzip Hoffnung*, 3 vols. (*passim*), Suhrkamp Verlag (1959), Frankfurt am Main. Translated by John Maxwell. Available online at https://thecharnelhouse.org/wp-content/uploads/2018/05/Ernst-Bloch-On-Karl-Marx.pdf

Boehm, C., Barclay, H. B., Dentan, R. K., Dupre, J. D., M.-C., Hill, S. Kent, Knauft, B. M., Otterbein, K. F., & Rayner, S. (1993). Egalitarian behavior and reverse dominance hierarchy [and comments and reply]. *Current Anthropology*, *34*(3), 227–254.

Bookchin, M. (1968/1986a). The forms of freedom. In *Post-scarcity anarchism* (pp. 163–192). Montreal, Quebec: Black Rose Books. Available online at https://theanarchistlibrary.org/library/murray-bookchin-the-forms-of-freedom

Bookchin, M. (1982). *The ecology of freedom.* Cheshire Books.

Bookchin, M. (1986b). *Post-scarcity anarchism.* Montreal, Quebec: Black Rose Books. Second edition, with a new introduction. Available online at https://libcom.org/files/Post-Scarcity%20Anarchism%20-%20Murray%20Bookchin.pdf

Boyd, R., & Richerson, P. J. (2009). Culture and the evolution of human cooperation. *Philosophical Transactions of the Royal Society B: Biological Sciences*, *364*(1533), 3281–3288.

Breugelmans, S. M., Zeelenberg, M., Gilovich, T., Huang, W.-H., & Shani, Y. (2014). Generality and cultural variation in the experience of regret. *Emotion*, *14*, 1037–1048.

Brighouse, H. (2021). If you're a socialist you need the Real Utopias Project, whether you like it or not. In *Paper Presented at the Conference "Engaging Erik Wright: Between Class Analysis and Real Utopias," October 9, 2021*.

Brodsky, I. (1991). *Poems* (p. 185). Eesti Raamat.

Brown, N., Lerer, A., Gross, S., & Sandholm, T. (2019). Deep counterfactual regret minimization. In *Proceedings of the 36th International Conference on Machine Learning*, Long Beach, CA.

Burns, T. R., & Engdahl, E. (1998). The social construction of consciousness. Part 2: Individual selves, self-awareness, and reflectivity. *Journal of Consciousness Studies*, *5*(2), 166–184.

Carr, D. (2021). Personal identity is social identity. *Phenomenology and the Cognitive Sciences*, *20*, 341–351.

Chakrabarti, A. (2018). The subject *is* freedom. *Philosophy East & West*, *68*(1), 277–297. Review of *Free Will, Agency, and Selfhood in Indian Philosophy*. Edited by Matthew R. Dasti and Edwin F. Bryant. New York: Oxford University Press, 2014.

Chatterton, P., & Pusey, A. (2020). Beyond capitalist enclosure, commodification and alienation: Postcapitalist praxis as commons, social production and useful doing. *Progress in Human Geography*, *44*(1), 27–48.

Chetty, R., Grusky, D., Hell, M., Hendren, N., Manduca, R., & Narang, J. (2017). The fading American dream: Trends in absolute income mobility since 1940. *Science*, *356*, 398–406.

Ci, J. (1999). Justice, freedom, and the moral bounds of capitalism. *Social Theory and Practice*, *25*(3), 409–438.

Clastres, P. (1987). *Society Against the State: Essays in Political Anthropology*. New York, NY: Zone Books.

Colson, D. (2011). Anarchism and the brute: Frank Norris, Herbert Spencer, and anti-government atavism. *Studies in American Naturalism*, *6*(1), 25–48.

Curry, O. S., Mullins, D. A., & Whitehouse, H. (2019). Is it good to cooperate? Testing the theory of morality-as-cooperation in 60 societies. *Current Anthropology*, *60*, 1.

Dale, R., Palma-Jacinto, S., Marshall-Pescini, S., & Range, F. (2019). Wolves, but not dogs, are prosocial in a touch screen task. *PLoS One*, *14*(5), e0215444.

Darwin, C. (1871). *The descent of man and selection in relation to sex.* London, UK: John Murray.

Davidai, S., & Gilovich, T. (2018). The ideal road not taken: The self-discrepancies involved in people's most enduring regrets. *Emotion*, *18*, 439–452.

Dean, H. (2016). Eudaimonia and 'species being': A Marxist perspective. In J. Vittersø (Ed.), *Handbook of Eudaimonic Well-Being*, Chapter 34 (pp. 507–520). Cham: Springer.

Digonnet, M. (2016). *Hiking Death Valley: A guide to its natural wonders and mining past* (2nd ed.). Quality Books.

Dilevko, J. (2014). The corporate state, technology, and the demise of morality. *Journal of Information Ethics*, *23*(2), 76–82. Review of C. Boehm, Moral Origins: The Evolution of Virtue, Altruism, and Shame. New York: Basic Books, 2012.

Dobzhansky, T. (1964). Biology, molecular and organismic. *American Zoologist*, *4*(4), 443–452.

Doctorow, C. (2017). *Walkaway*. New York, NY: Tor Books.

Easterlin, R. A., Morgan, R., Switek, M., & Wang, F. (2012). China's life satisfaction, 1990-2010. *Proceedings of the National Academy of Science*, *109*, 9775–9780.

Edelman, S. (2008). *Computing the mind: How the mind really works.* New York, NY: Oxford University Press.

Edelman, S. (2012). *The happiness of pursuit.* New York, NY: Basic Books.

Edelman, S. (2020). *Life, death, and other inconvenient truths.* Cambridge, MA: MIT Press.

Elangovan, A. R., & Hoffman, A. J. (2019). The pursuit of success in academia: Plato's ghost asks "What then?". *Journal of Management Inquiry*, *30*(1), 68–73.

Fairlie, S. (2009). A short history of enclosure in Britain. *The Land*, *7*, 16–31.

Fargher, L. F., Heredia Espinoza, V. Y., & Blanton, R. E. (2011). Alternative pathways to power in late Postclassic Highland Mesoamerica. *Journal of Anthropological Archaeology*, *30*, 306–326.

Fekete, T., van Leeuwen, C., & Edelman, S. (2016). System, subsystem, hive: boundary problems in computational theories of consciousness. *Frontiers in Psychology*, *7*, 1041.

Ferretti, F. (2017). Evolution and revolution: Anarchist geographies, modernity and poststructuralism. *Environment and Planning D: Society and Space*, *35*(5), 893–912.

Fields, C., Friston, K., Glazebrook, J. F., & Levin, M. (2022). A free energy principle for generic quantum systems. *Progress in Biophysics and Molecular Biology*, *173*, 36–59. Elsevier.

Freire, P. (1970). *Pedagogy of the oppressed.* New York: Continuum. English translation 2011.

Freire, P. (1974). *Education for critical consciousness.* New York: Continuum.

Friston, K. (2019). *A free energy principle for a particular physics.* Available online at https://arxiv.org/abs/1906.10184.

Frith, C. D., & Metzinger, T. (2016). What's the use of consciousness? How the stab of conscience made us really conscious. In A. K. Engel, K. J. Friston, & D. Kragic (Eds.), *The pragmatic turn: Toward action-oriented views in cognitive science*, Volume 18 of *Strüngmann Forum Reports* (pp. 197–224). Cambridge, MA: MIT Press.

Fung, Y.-L. (1966). *A Short History of Chinese Philosophy.* New York: Free Press. First published by Macmillan (1948). Edited by Derk Bodde.

Gintis, H., van Schaik, C., & Boehm, C. (2015). *Zoon Politikon*: the evolutionary origins of human political systems. *Current Anthropology*, *56*, 327–353.

Goldstein, D. M. (2010). Toward a critical anthropology of security. *Current Anthropology*, *51*(4), 487–517.

Graeber, D. (1996). Beads and money: notes toward a theory of wealth and power. *American Ethnologist*, *23*, 4–24.

Graeber, D. (2004). *Fragments of an anarchist anthropology.* Chicago, IL: Prickly Paradigm Press. Available online at https://www.prickly-paradigm.com/titles/fragments-anarchist-anthropology.html.

Graeber, D. (2006). Turning modes of production inside out or, Why capitalism is a transformation of slavery. *Critique of Anthropology*, *26*(1), 61–85.

Graeber, D. (2007). *There never was a West, or, Democracy emerges from the spaces in between* (pp. 329–374). Oakland, CA: AK Press. Available online at https://theanarchistlibrary.org/library/david-graeber-possibilities.

Graeber, D. (2020). Policy for the future of work. In R. Skidelsky & N. Craig (Eds.), *Work in the future: The automation revolution*, Chapter 16 (pp. 157–173). London: Palgrave Macmillan.

Graeber, D., & Wengrow, D. (2018, 2 March). How to change the course of human history (at least, the part that's already happened). *Eurozine.* Available online at https://www.eurozine.com/change-course-human-history/

Graeber, D., & Wengrow, D. (2021). *The dawn of everything: A new history of humanity*. London: Macmillan.

Gran, P. (2007). Modern world history as the rise of the rich: A new paradigm. *History Compass*, *5*(3), 1026–1049.

Harman, C. (2010). *Zombie capitalism: Global crisis and the relevance of Marx*. Chicago, IL: Haymarket Books.

Harris, Z., & Ryan, N. (2004). The inconsistencies of history: Vikings and Rurik. *New Zealand Slavonic Journal*, *38*, 105–130.

Henrich, J., & Muthukrishna, M. (2021). The origins and psychology of human cooperation. *Annual Review of Psychology*, *72*(24), 1–34.

Hickel, J. (2019). Degrowth: a theory of radical abundance. *Real-world Economics Review*, *87*, 54–68. Published 19 March 2019. Available online at http://www.paecon.net/PAEReview/issue87/Hickel87.pdf

Hoerl, C., & McCormack, T. (2016). Making decisions about the future: regret and the cognitive function of episodic memory. In K. Michaelian, S. B. Klein, & K. Szpunar (Eds.), *Seeing the future: Theoretical perspectives on future-oriented mental time travel*, Chapter 12 (pp. 241–266). Oxford: Oxford University Press.

Horkheimer, M., & Adorno, T. W. (2002). *Dialectic of enlightenment*. Stanford, CA: Stanford University Press. Originally published in 1944. Edited by Gunzelin Schmid Noerr. Translated by Edmund Jephcott.

House, B. R., Kanngiesser, P., Barrett, H. C., Yilmaz, S., Smith, A. M., Sebastian-Enesco, C., Erut, A., & Silk, J. B. (2020). Social norms and cultural diversity in the development of third-party punishment. *Proceedings of the Royal Society of London B*, *287*(1925), 20192794.

Houser, N., & Kloesel, C. (1992–98). The Peirce edition project. The Essential Peirce: Selected Philosophical Writings, 2.

Ichikawa, H. (1963). A preliminary conception of Zen social ethics. *Studies on Buddhism in India*, *11*(1), 359–348.

Ichikawa, H. (1965). Preliminary conception of Zen social ethics (continued). *Studies on Buddhism in India*, *13*(2), 844–836.

Kanngiesser, P., Schäfer, M., Herrmann, E., Zeidler, H., Haun, D., & Tomasello, M. (2022). Children across societies enforce conventional norms but in culturally variable ways. *Proceedings of the National Academy of Science*, *119*(1), e2112521118.

Kanu, I. A. (2017). Igwebuike as a hermeneutic of personal autonomy in African ontology. *Journal of African Traditional Religion and Philosophy (JATREP)*, *1*(2), 11–19.

Karpiak, K. G., & W. Garriott (2018). Introduction: Disciplines, fields, and problems. In K. G. Karpiak & W. Garriott (Eds.), *The anthropology of police*, Chapter 1 (pp. 1–20). Abingdon: Routledge.

Kropotkin, P. A. (1902). *Mutual aid: A factor of evolution.* London: W. Heinemann.

Laland, K. N., Uller, T., Feldman, M. W., Sterelny, K., Müller, G. B., Moczek, A., Jablonka, E., & Odling-Smee, J. (2015). The extended evolutionary synthesis: its structure, assumptions and predictions. *Proceedings of the Royal Society B: Biological Sciences*, *282*, 20151019.

Lane, R. (2009). Persons, signs, animals: a Peircean account of personhood. *Transactions of the Charles S. Peirce Society*, *45*, 1–26.

Le Guin, U. K. (2008). *Lavinia.* Boston: Mariner Books.

Lee, D. (1963). Freedom and social constraint. In D. Bidney (Ed.), *The concept of freedom in anthropology*, *Studies in General Anthropology* (Vol. 1, pp. 61–73). The Hague: De Gruyter Mouton.

Lenin, V. I. (1917). *The state and revolution.* Collected Works (Vol. 25, pp. 381–492). Available online at https://www.marxists.org/archive/lenin/works/1917/staterev/index.htm

Li, C. (2014). The Confucian conception of freedom. *Philosophy East and West*, *64*(4), 902–919. Special issue: Tenth East-West philosophers' conference, "Value and Values: Economics and Justice in an Age of Global Interdependence".

Lino e Silva, M., & Wardle, H. (2017). Introduction: Testing freedom. In M. Lino e Silva & H. Wardle (Eds.), *Freedom in practice: Governance, autonomy, and liberty in the everyday*, *Routledge studies in anthropology* (pp. 1–33). London: Routledge.

Marshall-Pescini, S., Schwarz, J. F., Kostelnik, I., Virányi, Z., & Range, F. (2017). Importance of a species' socioecology: Wolves outperform dogs in a conspecific cooperation task. *Proceedings of the National Academy of Science*, *114*(44), 11793–11798.

Marx, K. (1867/1887). *Capital: A critique of political economy.* Moscow: Progress Publishers. Translated by Samuel Moore and Edward Aveling, edited by Frederick Engels

Marx, K. (1961). The power of money. In *Economic and philosophic manuscripts of 1844*. Moscow: Foreign Languages Publishing House. Translated by Martin Milligan.

Metzinger, T. (2003). *Being no one: The self-model theory of subjectivity.* Cambridge, MA: MIT Press.

More, T. (1995). *Utopia.* Cambridge: Cambridge University Press. Originally published in Latin in 1516. Edited by George M. Logan, Robert M. Adams and Clarence H. Miller.

Morgan, J. (2018). Species being in the twenty-first century. *Review of Political Economy*, *30*(3), 377–395.

Mulgan, T. (2019). Corporate agency and possible futures. *Journal of Business Ethics*, *154*, 901–916.

Odouard, V. V., Smirnova, D. & Edelman, S. (2021). Polarize, catalyze, stabilize: conscience and the evolution of cooperation.

Oyeshile, O. A. (2007). Resolving the tension between corporate existence and the individual's freedom in African communal society: the Yoruba example. *Ultimate Reality and Meaning*, *30*(4), 278–300.

Riasanovsky, N. (1947). The Norman theory of the origin of the Russian state. *The Russian Review*, *7*, 96–110.

Robinson, W. I., & M. Barrera (2012). Global capitalism and twenty-first century fascism: a US case study. *Race & Class*, *53*(3), 4–29.

Sarkar, T. (2018). Commercial cash crop and the development of capitalist economy: A study of colonial tea plantations in Darjeeling hills. *Karatoya: NBU Journal of History*, *11*, 99–118.

Sawyer, J., & A. Gampa (2020). Work alienation and its gravediggers: Social class, class consciousness, and activism. *Journal of Social and Political Psychology*, *8*(1), 198–219.

Scott, J. C. (1998). *Seeing like a state: How certain schemes to improve the human condition have failed.* New Haven, CT: Yale University Press.

Servigne, P., & Chapelle, G. (2022). *Mutual aid: The other law of the jungle.* Polity. Translated by Andrew Brown.

Shell, M. (1989). The ring of Gyges. *Mississippi Review*, *17*(1/2), 21–84.

Sinclair, P. R. (1976). Fascism and crisis in capitalist society. *New German Critique*, *9*(Autumn), 87–112.

Souleimanov, E. A., Abbasov, N., & Siroky, D. S. (2019). Frankenstein in Grozny: vertical and horizontal cracks in the foundation of Kadyrov's rule. *Asia Europe Journal*, *17*, 87–103.

Spears, I. M. (2018). *Historical text and historical archaeology: Hunting the narrative of Kievan Rus.* Master's thesis, Eastern Kentucky University.

Su, Q., McAvoy, A., Mori, Y., & Plotkin, J. B. (2022). Evolution of prosocial behaviours in multilayer populations. *Nature Human Behaviour*, *6*, 338–348.

Valverde, M. (1996). The dialectic of the familiar and the unfamiliar: 'the jungle' in early slum travel writing. *Sociology*, *30*(3), 493–509.

Voland, E. (2014). The biological evolution of conscience — from parent-offspring conflict to morality. *Anthropological Review*, *77*, 251–271.

Walters, E. T., & de C. Williams, A. C. (2019). Evolution of mechanisms and behaviour important for pain. *Philosophical Transactions of the Royal Society B*, *374*, 20190275.

Ward, C. (2004). *Anarchism: A very short introduction.* New York, NY: Oxford University Press. Available online at https://theanarchistlibrary.org/mirror/c/cw/colin-ward-anarchism-a-very-short-introduction.a4.pdf.

Wilson, D. S., & Wilson, E. O. (2007). Rethinking the theoretical foundation of sociobiology. *The Quarterly Review of Biology*, *82*(4), 327–348.

Wither, J. K. (2020). Outsourcing warfare: Proxy forces in contemporary armed conflicts. *Security and Defence Quarterly*, *4*, 17–34.

Woodmass, K., & O'Connor, B. P. (2018). What is the opposite of psychopathy? A statistical and graphical exploration of the psychopathy continuum. *Personality and Individual Differences*, *131*, 254–260.

Wright, E. O. (2013). Transforming capitalism through real utopias. *American Sociological Review*, *78*, 1–25.

Wu, G. (2006). On the idea of freedom and its rejection in Chinese thought and institutions. *Asian Philosophy*, *16*(3), 219–235.

Yeats, W. B. (1966). *Selected poetry*. London: Macmillan. Edited by A. Norman Jeffares.

Zeelenberg, M., van den Bos, K., van Dijk, E., & Pieters, R. (2002). The inaction effect in the psychology of regret. *Journal of Personality and Social Psychology*, *82*, 314–327.

Part II

The Roads to Freedom

Interlude

At a crossroads.

Soyez réalistes, demandez l'impossible
(Be realist, demand the impossible)

— a May '68 slogan

What the sixties should teach us, then, is that there is no substitute for consciousness.

Post-Scarcity Anarchism
— Murray Bookchin (1986, p.50)

No sooner had people had a spare moment to reflect on their existence than they started groping around for remedies to being fully conscious at all times. It sure is fun to be completely there for the happy turns, but how is one supposed to carry on when not busy smelling a flower, having sex, or winning the lottery? The mix of anxiety, yearning, envy, guilt, remorse, and the rest of the self's evolution-tested and tried stimulants for self-improvement is a burden that is often too heavy to shoulder — and too depressing. Evolution appreciates it when that which does not kill you makes you stronger, but there is no reason for it to keep you happy in the process, so it doesn't.

As the hall monitor in the school of life, the conscious self is in charge of enforcing class attendance, even if it cannot ensure that all lessons are actually taken to heart. This is too bad, because the school district that sets the curriculum is a socioeconomic complex that is constructed and sustained by all, yet controlled by a select few. And so it is that the course of studies causes much anxiety and alienation: for most of the people, most of the time, life's main lesson is "Don't be poor."

But even in an ideal school, where the students are conscientious, the teachers kind and wise, and the lessons engrossing, uninterrupted instruction cannot be enjoyable, or effective. A good life too is sometimes hard to live. What to do?

One option is to try to catch some dreamless sleep, as it makes the self, and with it suffering, literally disappear. That, however, means that *you* do not get to be around to enjoy the relief. A perennially popular alternative is to play with the settings on your virtual reality rig, your brain. No need to open its casing, as it will void the warranty — everybody knows there's a variety of readily available chemicals that can temporarily take the edge off existence. In the meanwhile, though, the self that goes on existing is not quite the original *you*. Another resort is to eschew chemicals and instead mess with the settings

S. Edelman, *The Consciousness Revolutions*,
https://doi.org/10.1007/978-3-031-24012-6

by downloading the mind/brain owner's manual and learning from it some tricks, including the ultimate trick of letting go of the self while remaining awake.

These roads to freedom, such as it is, have two things in common. The first one is the cost they exact, which is either metamorphosis or oblivion: to walk these roads, you must change into someone else, or else into no one. The second one is the condition they impose, which is solitude: on these roads, you walk alone. There is, however, another possibility, another road.

This road is for people who refuse to be divided against each other by the powers that be. It too changes the walkers, not the least because it brings them together rather than keeping them apart, but also because this revolutionary road leads to a place where life can be made different. Daily existence will never be entirely without angst or sorrow, but wouldn't it be nice, for a change, not to have to worry about putting food on the table, or paying a doctor's bill, or making rent, or dying in a war over land, or water, or oil, or export markets? In the new place that lies just over the hills to the west, another world is possible.

❧ ❧ ❧

Reference

Bookchin, M. (1986). *Post-Scarcity Anarchism.* Montreal, Quebec: Black Rose Books. Second edition, with a new introduction. Available online at https://libcom.org/files/Post-Scarcity%20Anarchism%20-%20Murray%20Bookchin.pdf

6 Self-care

Suffering: eliminable and not. Too much self, too little social. The dance of life. The heart of a heartless world. Being and nothingness and zombies. The science of getting help. The sixth revolution.

> You remember the harbor. There were plenty of nice things floating around in it. That was the only time in my life I got so I dreamed about things. You didn't mind the women who were having babies as you did those with the dead ones. They had them all right. Surprising how few of them died. You just covered them over with something and let them go to it. They'd always pick out the darkest place in the hold to have them. None of them minded anything once they got off the pier.
>
> *On the Quai at Smyrna*
> — ERNEST HEMINGWAY

> What is to be done? We who are still half alive, living in the often fibrillating heartland of a senescent capitalism — can we do more than reflect the decay around and within us? Can we do more than sing our sad and bitter songs of disillusion and defeat?
>
> *The Politics of Experience*
> — R. D. LAING[1]

Suffering: Eliminable and Not

NO OTHER human pursuit demonstrates the concept of *eliminable* suffering as arrestingly as war. The true events behind Hemingway's story *On the Quai at Smyrna* happened as the Greco-Turkish war of 1919-1922 was winding down. Hundreds of thousands of Greek and Armenian non-combatants, caught in the ethnic cleansing campaign undertaken by Kemal Atatürk's army when it took over Izmir, perished in that conflict.[2] A century later, all the people who have experienced those events, with the possible exception of a centenarian or two among the babies born to the Greek refugee women on the quai, are long dead.

Death closes all — but not for those who saw their world burn and have to live with their memories of it and of how it ended. The Greek refugees from Anatolia, who settled in Piraeus and other poor neighborhoods of Athens, brought with them *rebetiko* — songs of existential sorrow, whose spirit has much in common with musical traditions such as the American blues, the Portuguese *fados*, and the Spanish flamenco.[3] What all these songs are about is

S. Edelman, *The Consciousness Revolutions*,
https://doi.org/10.1007/978-3-031-24012-6_6

REBETIKA: two songs from 1930s Greece, intended to illustrate the two categories of suffering. The exclamation *aman* means roughly "woe is me" or "mercy."[6]

Aman, aman A man must think, aman aman. . . at the hour of his death, aman, aman. . . ! that he'll go into the black earth, aman,. . . into the black earth, and his name will be erased, aman!	Aaaaaaaaaaa, Aaaaaaaaaaa, Aman! When a man grows poor, aman, aman, Aman! A man, it's a great sorrow. Aaaaaah! Aman, aman! It's a great sorrow *mete kemile* Aman, aman! He waits for death, aman, he waits because he has no courage. Aaaaman!

suffering — a universally familiar phenomenal experience that can be usefully defined (per chapter 3) as "consciously realized unhappiness of the inescapable variety."

Evolution has set us up with a capacity for being unhappy for an effectively unlimited range of reasons, so we should not expect suffering to be a monolithic category. Indeed, there are likely as many varieties of it as there are sentient beings. Still, specifically human suffering always belongs to either of two kinds: that which is eliminable and that which is not. These are represented here by the lyrics of two *rebetika* songs: one about the inevitability of death, the other about such poverty as makes you wish you were dead.

Because only conscious beings *can* suffer, and because all naturally evolved conscious beings *do* in fact suffer on many occasions during their lifetimes, suffering is an integral quality of conscious existence for creatures like us.[4] This vintage observation, which has over millennia provided thought fodder for philosophers and endless plot ideas for tragedy writers,[5] we may actually be able to elaborate a bit on. Rather than blame suffering on consciousness indiscriminately, we can tell that it is promoted by a kind of imbalance between complementary types of consciousness: self-consciousness, of which there is too much; and social consciousness, of which there is too little.

Too Much Self, Too Little Social

Let's start with self-consciousness. Why would having too much of it amplify suffering? Because suffering is predicated in the first place on the ongoing functioning of the self. The chores of the computational processes that comprise the self include maintaining the sense of ownership, not just of the self's physical body (as discussed in chapter 3), but also of its thoughts and the ever-present feelings. According to the best available theory of suffering, due to Thomas Metzinger, when some of these feelings are negative and the state of affairs that induces them cannot be immediately eluded, suffering ensues.[7] In contrast, when for some reason the self happens not to be in operation, negative affect can still occur (and even help steer behavior), but suffering cannot, because there is no one to experience it.

That would be pretty much the complete story of human suffering, were we not an ultrasocial species. Why would people get together to play and listen to sad music and to each other's complaints about the hard life, if doing so were not helping them bear it? By feeding the sense of "being in this together," a well-developed social consciousness acts to mitigate suffering both on the group and the personal levels; those who feel, or are made to feel, as if they don't belong are bound to suffer more.[8] On the utilitarian account, however, the worst (and likely the worst-off) are those in whom social consciousness is not merely lacking but is corrupted by extreme individualism, misanthropy, or sociopathy. In addition to suffering more themselves, such people, through action or inaction, cause extra suffering among others, in proportion to the power they wield.

The Dance of Life

The manner in which social consciousness can help with suffering can be appreciated if we consider how it works in a literally hopeless case: the suffering induced by the peculiarly and essentially human insight into personal mortality. Once this knowledge has been attained, it cannot be unlearned. The resulting state of mind would thus appear to epitomize the category of unavoidable suffering. (Curiously, the opposite fate, personal immortality, seems to terrify people just as thoroughly, once they grasp its implications.[9])

In facing death, a self-conscious being experiences an irresolvable tension between, on the one hand, its built-in bias in favor of continued existence, which plays a part in the process of self-evidencing (described in chapter 3), and on the other hand the impending certain end to that existence. We, who are capable of envisaging and ruminating on the future, are primed to experience this tension the first time the realization drops that we are bound eventually to share the fate of our pet hamster or grandmother. From that point on, pretense and self-distraction are the order of the day, every day: as Peter Zapffe notes, "Most people learn to save themselves by artificially limiting the content of consciousness." [10]

In the beast, suffering is self-confined, in man, it knocks holes into a fear of the world and a despair of life.
— P. W. Zapffe (1933)

I find it amazing how social consciousness can step in here to blunt the existential pain, not by suppressing one's awareness of the sorry state of the world, but by elevating it. Singing about death — and at times dancing with death personified, as in the medieval European tradition of *Danse macabre*, or the Dance of Death[11] — clearly helps, or we wouldn't be so into it.[12] By turning her soul inside out, the *rebetika* singer creates a shared space in which we can face, without flinching, life, and even death.

The Heart of a Heartless World

The spirit of *rebetika*, which is an entirely secular phenomenon, is nevertheless perfectly captured by the language used by Marx to describe *religion* — "the sigh of the oppressed creature" — in the sentence that immediately precedes the much more famous "opium of the masses" phrase. In that, *rebetika* and other songs of suffering differ radically from protest songs and revolutionary anthems. A great example of the latter is the *Warszawianka* (translated into

Religion is the sigh of the oppressed creature, the heart of a heartless world, and the soul of soulless conditions. It is the opium of the people.
— Karl Marx (1970)

English as *Whirlwinds of Danger*), which dates to the Polish Revolution of 1905. Immediately upon mentioning the workers' condition, the song dares the listener to ignore it:

> Brothers and sisters in hunger are calling,
> Shall we be silent to their sorrow and woe?

The very next stanza is a call for action:

> Off with the crown of the tyrants of evil!
> Down in the dust with the prince and the peer!

In contrast, the songs of suffering do not entreat the listener to take action, nor do they propose any. Instead, they present the "wretched of the world," who already know what they are, with a mirror that reflects back their condition.

This sustained and exclusive focus on personal suffering may seem perverse — even reactionary, insofar as it draws attention away from systemic injustice — yet it makes a certain sense: to think clearly about the causes behind one's plight, let alone to be able to plan and organize for a systemic intervention, one must first gain some breathing room. This is why any project that aims at understanding and assuaging human suffering must begin with a look at societal institutions that focus on its personal dimensions. Religion, "the heart of a heartless world," is one such institution.

Even in its mundane — as opposed to millennarian[13] — manifestations, religion can serve as an effective personal palliative of suffering caused by poverty and powerlessness. Indeed, empirical studies suggest that religiosity eases the psychological burden of poverty and is associated with higher subjective well-being.[14]

These effects are fully expected in the case of Christianity, which has apotheosized suffering unto death, made poverty a virtue, and promised the poor eternal bliss in the company of a crucified and resurrected Son of God. What better solace can there be for the faithful than the knowledge that their suffering earns them the compassion of the Lady of Sorrows? —

> Santa Maria das dores, Mãe de Deus, se for pecado
> Tocar e cantar o fado, rogai por nós pecadores
> Nenhum fadista tem sorte, rogai por nós Virgem Mãe
> Agora, sempre e também na hora da nossa morte
>
> Holy Mary of Sorrows, Mother of God, if it is a sin
> To play and sing *fado*, pray for us sinners
> No fado singer is lucky,[15] pray for us Virgin Mother
> Now, always and also at the hour of our death

Are the rich barred from the kingdom of heaven? On the one hand, we read: "Then said Jesus unto his disciples, Verily I say unto you, That a rich man shall hardly enter into the kingdom of heaven. And again I say unto you, it is easier for a camel to go through the eye of a needle, than for a rich man to enter into the kingdom of God" (Matthew 19:23-24). On the other hand, the very next verse walks that notion right back: "When his disciples heard it, they were exceedingly amazed, saying, Who then can be saved? But Jesus beheld them, and said unto them, With men this is impossible; but with God all things are possible" (Matthew 19:25-26).

Those among the religious solace seekers who decide to take scripture too literally are, however, liable to have their faith shaken, and peace of mind disturbed, by a suspicion that they are being led on. On the one hand, there is the promise that the poor person's lifetime of suffering will be made up for in the hereafter, while the rich are left waiting for God to first put a camel through the eye of a needle. On the other hand, it would seem that God, with whom "all things are possible," could have abolished poverty in the first place.

What the questioning believer faces here is, of course, the problem of *theodicy*: how to account for the presence of evil in a universe presided over by a just and omnipotent deity. Whether or not it is possible to find a solution to it within a theological framework that presupposes that the presiding deity is just and omnipotent is beside the point if we'd rather seek to improve the human condition than be offered an excuse for it, no matter how authoritative.[16]

Philosophical theology aside, then, one may be moved to adopt a working hypothesis, according to which the continued existence of eliminable suffering is down to worldly powers. As these are likely to be the very same rich people whom the Gospel (sometimes) scorns, it would seem that a cognitive dissonance between faith and facts is inevitable. Still, our ability to compartmentalize our beliefs and political-economic convictions ensures that a class perspective on suffering does not necessarily endanger the religiosity of those who develop it in themselves.

Indeed, studies in the sociology of religion consistently demonstrate that lower socioeconomic status is typically associated with higher religiosity, apparently because religion serves as an effective psychological defense against despair.[17] That said, higher economic inequality reportedly boosts religiosity across all income levels — a finding that indicates that both the poor and the rich practice system justification, a secular variety of theodicy.[18]

In any case, whether or not Christian dogma can be reconciled with class consciousness — as attempted by the 1970s "liberation theology" movement in Latin America[19] — for the abatement of suffering it is important that both the faith-based and the class-analysis mindsets remain available to the individual. This is because, just as social existence determines consciousness, it shapes the way religion is interpreted in daily life, and not vice versa. In particular, as Ruth Benedict has observed,[20]

> [T]he will to righteousness has to be inculcated in the daily round of life or it will not find expression in religion. In their religions people have cast on a screen their hates and maliciousnesses if that was what their society inculcated, or warmth and righteousness if *that* was what was possible in their daily lives.

Societies with high synergy construct religions in which they pray and dance and sing for benefits that are benefits to all the tribe. ... Tribes with low synergy, however, with little experience in their daily lives of any benefits not gained at the expense of others, typically build their religious practices upon these privately owned supernatural powers and use them according to the habits of aggression set up in their daily lives.
— Ruth Benedict (1970, p.331)

Rather than merely reconciling religious and class consciousness, one might seek for *synergy* between them, which can serve as an even more potent catalyst of social justice. For a role model in this matter, we can look to Martin Luther King, Jr.: if every religious leader of his caliber were as socialist as he was, the world would be a better place.[21] And of course, the people's faith in the possibility of such a world of justice for all has found its way into folk songs, such as this Black spiritual, which I like best as performed by Paul Robeson, a lifelong socialist, and the deepest bass:

> Didn't my Lord deliver Daniel?
> And why not every man?

Being and Nothingness and Zombies

In the absence of a clear sign from on high that the deliverance of "every man" is forthcoming, it is quite natural for people to seek palliatives for personal suffering that do not depend on the good will of a reluctant and possibly nonexistent supreme being. The most radical palliative of this kind is offered by Buddhism: a finely honed philosophical theory of selfhood and suffering, conveniently disguised as religion for reasons of broader appeal and general synergy.[22]

The supremely analytical Buddhist doctrine rests on the proclamation of Four Noble Truths, of which the first is the inherence of *dukkha*, or suffering, in the existence of all sentient beings; and the last is the availability of a Noble Eightfold Path to *moksha*, or liberation from suffering (as well as from the cycle of death and rebirth, a supernatural twist, inherited from Hinduism, which is tangential to the logical foundations of Buddhism as applied philosophy).

Earlier in this book, I have discussed a key Buddhist insight into consciousness, as expressed in the doctrine of dependent origination — an observation of the all-encompassing nature of the great web of cause and effect, and a foundation of the other central doctrine of Buddhism, *śūnyatā* or emptiness (that is, everything's emptiness of independent existence). In contemplating the problem of personal suffering, we can now revisit parallels between the theoretical picture painted by the Buddhist philosophy of the self and its computational understanding. First, as noted in chapter 3, we must remember that what the well-known *anatta* or no-self doctrine posits the non-existence of is an essential, independent, and permanent self. This notion fits well our emerging view of the self as an ephemeral bundle of computational processes that are, furthermore, subject to certain conditions.

On Metzinger's theory (mentioned earlier in this chapter; see note 7 on p.174), abolishing some of these conditions would amount to abolishing suffering. The first candidate is consciousness itself: where there is no phenomenal experience, there can be no suffering. Thus, the mythological p-zombies, had they existed, would be fundamentally incapable of it. In any case, this intervention is too extreme: permanently extinguishing consciousness — as in attaining *nirvana* (literally, a "blowing out") — is effectively equivalent to suicide, a price that most people deem too high for the cessation of suffering.

Metzinger's second and third conditions for suffering — respectively, the self's ownership of all feelings and the negativity of some of them — present an opportunity for a less drastic intervention. Simply refraining from negative emotions, even if it can be managed, is not a feasible option, because any effort to artificially exclude negatively valenced affective states would result in a swift recalibration of the range of affect, returning the sufferer back to square one. Might giving up on the ownership of all affect work? This recourse aligns with the Buddhist precept of *upekkha* or equanimity. If successful, it turns the former sufferer, if not into a complete zombie, then at least into a creature whose range of conscious states is qualitatively very different from that of a typical human.

The fourth condition that enables suffering likewise echoes Buddhist doctrine. It has to do with transparency — a key property of phenomenal states, which came up briefly on p.75, at the end of chapter 3, where I discussed the evolution of suffering. Conscious states are transparent in the sense that they are representations (of states of affairs elsewhere in the system, or outside it), yet are not perceived as such. Attaining an insight into this condition (presumably, through proper training in meditation; an intellectual understanding does not seem to suffice) is the first step towards liberation from the tyranny of mere representations. A further realization of the representational nature of the conscious self, and its "dependent origination" with that which it represents, completes the move. In the emptiness that remains, there is no form, no feeling, no perception, no volition, no consciousness — and no suffering.[23]

It would seem that the principle of dependent origination, which is absolutely central to Buddhism, might predispose it doctrinally to social engagement. If my well-being depends on how others fare, and vice versa, shouldn't we all try to be a bit nicer to each other, as the Universe expects of us? That, however, has not happened: despite calls for the development of an "engaged Buddhism,"[24] it remains an intensely personal path to walk. Indeed, the Buddhist notion of "taking refuge" in the *sangha* or the community is given in the *Dhammapada* a rather startling interpretation: "One indeed is one's own refuge; how can others be a refuge to one?"[25] The "community" in question is therefore nothing more than an abstract designation for the set of individual seekers after enlightenment, each of whom is urged to turn away from the world and all its inhabitants:[26]

> Having let go of passion,
> aversion,
> delusion;
> having shattered the fetters;
> unfazed at the ending of life,
> wander alone
> like a rhinoceros.

The Science of Getting Help

As we saw so far in this chapter, there are many things you can do to make your condition a little more bearable — and whose condition could not use even just a little bit of improvement? You can sit in a cafe that plays sad music and talk with friends about life. You can stop by a church and pray for mercy, now and at the hour of your death. You can get trained to let go of everything in life, including your very self, in the hope that the heavy load will disappear with it. Or you can get professional help.

It is the supposedly science-based interventions staged by mental health professionals that deserve our strongest mistrust — not because of the therapists' ineptitude or bad intentions, but because in our political-economic system the very conception of "mental health" is problematic. Under a system that is built on profound disregard for people's well-being and at the same time an expectation that everyone work hard and be happy, a person who pauses for a moment to contemplate the state of everyone's affairs is likely to become depressed and angry.[27] A mental health service based on the assumption that there's nothing basically wrong with the world would naturally diagnose such a person as not well — criminally so, if the person does not learn to conform and transgresses against the law of the land.

The absurdity of the present system of rule is so transparent to healthy consciousness that it needs sick consciousness to keep itself alive.
— Max Horkheimer and Theodor Adorno (2002, pp.163-164)

Medicalization and criminalization of suffering take different forms, depending on the prevailing politics. In an openly totalitarian state, such as the now defunct USSR or the "communist" China, a dissident can get forcibly committed to a mental health institution. In comparison, in a neoliberal corporate oligarchy such as the "democratic" West, dissidence, let alone mere depression, is ignored — unless the dissidents attempt something really subversive, such as feeding kids for free at school (an initiative that brought the wrath of the FBI down on the Black Panthers). At the same time, the corporate state peddles to the masses opium of all sorts, both literal (as in the case of the U.S. opioid scandal)[28] and metaphorical (popular entertainment, spectator sports, the political spectacle, state-sponsored religions).

Against this gloomy backdrop, the courage and the skills of those therapists who are aware of the nature of the system they are up against, and who are explicitly committed to social justice,[29] should be appreciated all the more. A good therapist knows how to walk the fine line between blaming too much on the system and failing to recognize the objective conditions it poses. They know that the best psychotherapy "takes as fundamental the social relations among persons."[30] And they know which of the available methods work better than others.

A loose label for one family of methods that do is cognitive behavioral therapy or CBT, which recruits the client's capacity for self-understanding to help them feel better — even if the objective contextual contributing factors cannot be removed.[31] CBT begins with acknowledging a human universal: the overarching desirability of self-knowledge. Further, a case can be made that CBT is essentially about self-growth, that is, achieving one's potential despite the difficulties. In this, it may be contrasted to radical interventions based on the Buddhist doctrine or on other theories of suffering, which aim to fundamentally alter the sufferer's conscious self. In comparison, scientifically informed and socially aware cognitive therapy not only preserves, but actually expands the client's consciousness — a small strike for the next revolution, one self at a time.

The Sixth Revolution

Each of the five consciousness revolutions that together got us to this point marked the emergence of a hitherto unheard-of type of consciousness: from minimal phenomenality, through selfhood and self-awareness, to social consciousness. The sixth one is different. Its key component — conscious insight into the ubiquity of suffering — grows out of a combination of the self's existing characteristics and capacities, notably, affective introspection and language-mediated thought. This revolution is also incomplete, in that its core insight lacks closure. Like Adam and Eve, those of us who learn about the suffering that pervades all existence must then live with that knowledge, but without a clear notion of what to do about it, while serving a life sentence away from Eden.

The refugees from Eden are pained, confused, and afraid. They do not know who to blame for their state. Some of them suspect that the police angel with the flaming sword who guards the little postern from which they have just been ejected is not on their side. Still, many are intent on appeasing the Big Man in charge and winning readmittance to the garden. Opinions on how best to do so differ widely. One thing seems to be clear: keeping oneself busy and distracted at all times works as well as anything in staving off the terrifying lucidity of full consciousness of one's condition. And so, exhausted from a long day of earning their living by the sweat of their brow, the refugees settle into an uneasy sleep, haunted by monsters of their own device. Will dawn never come?[32]

can't we awaken?
or are the forces of life greater than we are?
can't we awaken? must we forever,
dear friends, die in our sleep?

❧ ❧ ❧

Notes

1. Hemingway's short story, based on his coverage of the Greco-Turkish war for the *Toronto Star* in 1922, opens the collection *In Our Time* (Hemingway, 1930). The Laing (1967, p.11) quote is from the Introduction to *The Politics of Experience*.

2. Witnessing the atrocities of war in Anatolia may have been a formative experience for the 23-year old Hemingway (Long, 2019), but it was not his first one, because it was not the first war that he saw up close. Four years earlier, serving as a volunteer in the Great War (called by H. G. Wells "the war that will end war"), on the Italian Front, he had been seriously wounded and had to spend six months at a hospital. Hemingway's subsequent lifelong disdain for war and for the military brass who are insulated from its worst consequences was not, however, accompanied by any kind of political awakening.

3. Dalianoudi (2005) offers fascinating historical context for the emergence of *rebetika* and related musical styles. The psychology behind people's enjoyment of sad music has been given a book-length treatment by Garrido (2017).

4. The distinction between non-eliminable and eliminable suffering is illuminated in this passage from Ursula K. Le Guin's *The Dispossessed* (1974, ch.2, p.60):

> Suffering is the condition on which we live. And when it comes you know it. You know it as the truth. Of course it's right to cure diseases, to prevent hunger and injustice, as the social organism does. But no society can change the nature of its existence. We can't prevent suffering. This pain and that pain, yes, but not Pain. A society can only relieve social suffering — unnecessary suffering. The rest remains. The root, the reality.

Anderson (2015) is a snapshot of the statistics of life satisfaction and the prevalence of suffering all over the world. For further discussion, and some references, see the chapter on suffering in (Edelman, 2020, ch.32).

5. Readers interested in some entry points into the literature of suffering can start with these authors: Zapffe (1933), Bourdieu (1999), Erofeev (1992), Ligotti (2010), Parfit (2011), Roy (2015), Vinding (2020).

6. The translations of the two *rebetika* songs are reproduced with permission from Holst-Warhaft (2018, pp.95-96), who notes (personal communication): "As for the words *mete kemile*... they have no meaning in Greek or Turkish

that I can find. It is an exclamation and so far I haven't found any more information although I've hunted around. So I thought best to leave the phrase untranslated."

7. Metzinger's theory of suffering is laid out in (Metzinger, 2017).

8. There is evidence that social activities contribute to the mitigation of personal pain and suffering; this includes making music (Dunbar et al., 2012) and participating in nostalgia-related rituals (Sedikides and Wildschut, 2019).

9. I discuss the aversion to immortality (among other things) in *Identity, immortality, happiness: pick two* (Edelman, 2018).

10. In their review of the anthropology of death, Palgi and Abramovitch (1984) note that "Death awareness is a natural sequel to the development of self-awareness-an intrinsic attribute of humankind. The consciousness of man's transience in its known earthly form is thus a universal phenomenon." The consciousness-limiting move against existential suffering arising from one's awareness of death is mocked by Zapffe (1933, III):

> Here is, by the way, an opportunity for the wildest round-dancing through ever higher ironic levels, into a most embarrassing *circulus vitiosus*. Here one can chase one's ego across numerous habitats, enjoying the capacity of the various layers of consciousness to dispel one another.

In his book *The Conspiracy against the Human Race: A Contrivance of Horror* (2010), Ligotti (who happens to be a professional writer of "supernatural horror") extends the mockery to everything that religion, philosophy, and psychology ever had to say about suffering.

11. For a fascinating etymological, historical, and anthropological examination of the *Danse macabre*, see (Eisler, 1948). Gamliel (2004) focuses on how this concept plays out among the present-time elderly.

12. A 208-reference review of the biology, psychology, and culture of sad music (Eerola et al., 2018) paints a "sombre picture" of our (lack of) understanding of how it works.

13. Concerning the relationship between millennarian religious movements and their contemporary politics, James C. Scott (2009, p.294) notes that the two can hardly be separated:

> Given their elements of divine intervention and of magic, it is all too easy to exoticize prophetic movements cast in religious garb. This is a temptation that should be resisted. It should be resisted because virtually all popular struggles for power that today would qualify as "revolutionary" were, before the last quarter of the eighteenth century, generally understood in a religious idiom. Popular mass politics was religion, and religion was political. To paraphrase Marc Bloch, millennial revolt was as natural to the seigneurial (feudal) world as strikes, let us say, are to large-scale

> capitalism. ... Ideas of justice and of rights and, indeed, what we might today call 'class consciousness' were religiously phrased.

For a more focused treatment of the subject of church and revolution, see (Kowalewski and Greil, 1990).

14. According to Minkov et al. (2020, p.1), "Numerous studies have reported a positive individual-level association between happiness and two psychologically distinct states of mind: religious faith and subjective freedom (a feeling of life control)." For evidence that religiosity eases the perceived burden of poverty, see (Berkessel et al., 2021).

15. The invocation of luck (or rather lucklessness) in this Portuguese song, *Ave Maria Fadista*, reminds me of a phrase I came across in a book on the history of freedom (de Dijn, 2020, p.329) — "safety net for Americans down on their luck" — behind which there is the assumption that falling into poverty in the U.S. is a matter of some cosmic lottery, rather than systemic economic injustice.

16. Theodicy (Betenson, 2016; Mavelli, 2016) is arguably best approached as a problem not in theology but in social studies and praxis. For instance, Hamilton (2015) applies this concept to the "good Anthropocene" propaganda, which seeks to justify our ongoing destruction of the planetary ecosystem. Historically, it is interesting to contrast the theologies of Martin Luther and of Thomas Müntzer, the pastor who led the rebels in the German Peasant War of 1525 (Toutge, 2018). Unlike his much more famous contemporary, Müntzer joined workers and peasants in demanding a "kingdom of heaven on earth," the first step towards which would be to ensure livelihood for all; per Toutge (2018, p.19):

> Everyone should properly receive according to his need. Any prince, count, or lord who refuses to do this even when seriously warned should be hanged or have his head chopped off.

17. Religion's role as a psychological defense is documented, for instance, in (Brandt and Henry, 2012). Flannelly et al. (2006) found that in their subjects socioeconomic status was inversely related to religious attendance, prayer, and religious fundamentalism.

18. McCloud (2007) asks whether or not class leads individuals to be attracted to certain religions, answering with "a qualified yes" (p.851). At the same time, religion generally seems to hold a broad appeal in the U.S.: as Solt et al. (2011) reports, economic inequality has a strong positive effect on the religiosity of all members of a society regardless of income.

19. A quick introduction to liberation theology can be found in (King, 2016; Levine, 1988); McLaren and Jandrić (2018) offer a Marxist analysis. For a definitive treatment of liberation theology in the context of psychotherapy, see (Martín-Baró, 1996); I thank Nikolay Kharkov for sharing the syllabus for his Fall 2022 course *Political Psychotherapies*, in which I came across this and several other eye-opening readings.

20. Marx's famous materialist take on the relationship between existence and consciousness is found in his Preface to *A Contribution to the Critique of Political Economy* (1977):

> It is not the consciousness of men that determines their existence, but their social existence that determines their consciousness.

Maslow and Honigmann (1970, p.331) makes a similar connection between existence and religion:

> [I]n their religion, societies have transcribed and apotheosized the cooperativeness or the aggressions their cultural life arouses.

21. The socialist convictions of Martin Luther King, Jr. have been well-documented (e.g. Sturm, 1990).

22. See *Buddhism as Philosophy* (Siderits, 2007). A comparison of the conceptions of the self in Wittgenstein, Hume, and Buddhism can be found in (Richards, 1978).

23. This line is, of course, taken straight from the *Prajñāpāramitā-Hṛdayam Sūtra*, or the Great Heart Wisdom Sutra:

> *Tasmāc Śāriputra, śūnyatāyāṁ*
> *na rūpaṁ, na vedanā, na saṁjñā, na saṁskārāḥ, na vijñānam*
> (Therefore, Śāriputra, in emptiness
> there is no form, no feeling, no perception, no volition, no consciousness)

The nonexistence of the self, *anatta*, is terrifying to face — if viscerally experienced rather than intellectually contemplated (in the words of Laing (1967, p.35), " 'There's nothing to be afraid of.' The ultimate reassurance, and the ultimate terror."). In psychiatry, this experience can be classified as "oceanic self-dissolution" (see Metzinger, 2003, for a detailed discussion of this and other disorders of consciousness), which can be achieved by the administration of psylocybin (Lebedev et al., 2015). Pickering (2019) offers a Jungian interpretation of an integrative "case study" of the dire effects of self-dissolution on a Western Buddhist which, unfortunately, appears to be based on a serious misunderstanding of the concept of *śūnyatā*. An actual case study of the attitude to death in several religious groups, including lay and monastic Buddhists (Nichols et al., 2018) revealed that the latter experience the most death anxiety, in complete contradiction to the doctrine that they officially espouse.

24. Although the concept of engaged Buddhism has its own entry in a recently published encyclopedia of religion (Gleig, 2021), historically it appears to have been rather marginal among the Buddhist schools of thought. Attempts to develop it have been made by the Anarchist Zen monk Tàixū in the 1930s China and Taiwan (Dessein, 2020) and, somewhat more recently, by the Japanese anti-war philosopher Ichikawa (1963; 1965).

25. According to tradition, this verse (Dhammapada, 1986, 160) was uttered by the Buddha in response to the story of the mother of Kumarakassapa, a

bhikkuni who attained arahatship on the same day as she renounced all attachment to her son. A psychologist would point out in this connection that the human sense of affiliation is central to normal development and functioning through the lifetime (Feldman, 2020) and that destroying it is likely to have serious negative repercussions for the "liberated" person's mental well-being.

26. The rhinoceros verses are from the Khaggavisana Sutta (Sn 1.3 PTS: Sn 35-75), translated from the Pali by Thanissaro Bhikkhu (https://www.dhammatalks.org/suttas/KN/StNp/StNp1_3.html).

27. On the subject of the relationship between suffering and happiness, Metzinger (2017) argues that as a society we should normatively aim to reduce the former rather than increase the latter. A detailed proposal for suffering-focused ethics can be found in (Vinding, 2020).

28. Concerning medicalization, Brinkmann (2014, p.642) comments that "Viewing people's lives through a diagnostic lens de-politicizes their problems and turns them into a matter of personal health and illness." The Black Panthers' food program is described in (Potorti, 2017). For a glimpse of the racism and the other types of injustice perpetrated under the cover of the U.S. "war on drugs," see (Netherland and Hansen, 2017).

29. DeBlaere et al. (2019) place counseling in the context of social justice. Comas-Díaz (2020) describes "liberation psychotherapy," a concept inspired by the work of Ignacio Martín-Baró (1996).

30. The quote regarding the need for psychotherapy to acknowledge the primacy of social relations among people is from (Laing, 1967).

31. For empirical evidence in support of CBT, see (Fordham et al., 2021). Dijkstra and Nagatsu (2021) offer a comparison of several therapeutic techniques, including CBT. Interestingly, a mindfulness intervention has been found ineffective in similar contexts (Kaplan et al., 2022).

32. The lines that close this chapter are from Charles Bukowski's poem *and the sun wields mercy*, first printed in EPOS 12:2, p.6 (1960).

References

Anderson, R. E. (Ed.). (2015). *World suffering and quality of life*, Volume 56 of *Social indicators research series*. New York, NY: Springer.

Berkessel, J. B., Gebauer, J. E., Joshanloo, M., Bleidorn, W., Rentfrow, P. J., Potter, J., & Gosling, S. D. (2021). National religiosity eases the psychological burden of poverty. *Proceedings of the National Academy of Science, 118*(39), e2103913118.

Betenson, T. (2016). Anti-theodicy. *Philosophy Compass, 11*, 56–65.

Bourdieu, P. (1999). *The weight of the world: Social suffering in contemporary society*. Cambridge, UK: Polity.

Brandt, M. J., & Henry, P. J. (2012). Psychological defensiveness as a mechanism explaining the relationship between low socioeconomic status and religiosity. *International Journal for the Psychology of Religion*, *22*, 321–332.

Brinkmann, S. (2014). Languages of suffering. *Theory & Psychology*, *24*, 630–648.

Comas-Díaz, L. (2020). Liberation psychotherapy. In L. Comas-Díaz & E. Torres Rivera (Eds.), *Liberation psychology: Theory, method, practice, and social justice*, Chapter 9 (pp. 169–186). American Psychological Association. Available online at http://www.jstor.org/stable/j.ctv1chs1sn

Dalianoudi, R. (2005, September). Rebetiko song, tango, fado, flamenco & blues — historical context. *Journal of I.O.V. UNESCO*, *5*, 21–27.

de Dijn, A. (2020). *Freedom: An unruly history*. Cambridge, MA: Harvard University Press.

DeBlaere, C., Singh, A. A., Wilcox, M. M., Cokley, K. O., Delgado-Romero, E. A., Scalise, D. A., & Shawahin, L. (2019). Social justice in counseling psychology: then, now, and looking forward. *The Counseling Psychologist*, *47*(6), 938–962.

Dessein, B. (2020). The heritage of Taixu: Philosophy, Taiwan, and beyond. *Asian Studies*, *8*(3), 251–277. Special Issue: Taiwanese Philosophy and the Preservation of Chinese Philosophical Traditions / Modern Transformations in Logic, Daoist and Buddhist Philosophy.

Dhammapada. (1986). *The Dhammapada: Verses and stories*. Rangoon, Myanmar: Burma Tipitaka Association. Translated by Daw Mya Tin. Available online at https://www.tipitaka.net/tipitaka/dhp/index.php

Dijkstra, J. M., & Nagatsu, T. (2021). Cognitive behavioral therapy (CBT), acceptance and commitment therapy (ACT), and Morita therapy (MT); comparison of three established psychotherapies and possible common neural mechanisms of psychotherapies. *Journal of Neural Transmission*. https://doi.org/10.1007/s00702-021-02450-9

Dunbar, R. I. M., Kaskatis, K., MacDonald, I., & Barra, V. (2012). Performance of music elevates pain threshold and positive affect: implications for the evolutionary function of music. *Evolutionary Psychology*, *10*(4), 688–702.

Edelman, S. (2018). Identity, immortality, happiness: pick two. *Journal of Evolution and Technology*, *28*, 1–17.

Edelman, S. (2020). *Life, death, and other inconvenient truths*. Cambridge, MA: MIT Press.

Eerola, T., Vuoskoski, J. K., Peltola, H.-R., Putkinen, V., & Schäfer, K. (2018). An integrative review of the enjoyment of sadness associated with music. *Physics of Life Reviews*, *25*, 100–121.

Eisler, R. (1948). Danse macabre. *Traditio*, *6*, 187–225.

Erofeev, V. (1969/1992). *Moscow to the end of the line*. Chicago, IL: Northwestern University Press. Translated from Russian by H. W. Tjalsma and A. R. Dirik.

Feldman, R. (2020). What is resilience: An affiliative neuroscience approach. *World Psychiatry*, *19*, 132–150.

Flannelly, K. J., Koenig, H. G., Ellison, C. G., Galek, K., & Krause, N. (2006). Belief in life after death and mental health: Findings from a national survey. *Journal of Nervous & Mental Disease*, *194*, 524–529.

Fordham, B., Sugavanam, T., Edwards, K., Stallard, P., Howard, R., das Nair, R., Copsey, B., Lee, H., Howick, J., Hemming, K., & Lamb, S. E. (2021). The evidence for cognitive behavioural therapy in any condition, population or context: a meta-review of systematic reviews and panoramic meta-analysis. *Psychological Medicine*, *51*, 21–29.

Gamliel, T. (2004). The danse macabre: A unique discourse among the elderly. *Semiotica*, *149*, 261–275.

Garrido, S. (2017). *Why are we attracted to sad music?* London: Palgrave Macmillan.

Gleig, A. (2021). Engaged Buddhism. In *Religion*, *Oxford research Encyclopedias*. New York, NY: Oxford University Press.

Hamilton, C. (2015). The theodicy of the "Good Anthropocene". *Environmental Humanities*, *7*, 233–238.

Hemingway, E. (1930). On the Quai at Smyrna. In *In our time* (pp. 11–12). New York: Scribner.

Holst-Warhaft, G. (2018). *Amanes*: The legacy of the oriental mother. In T. Bucuvalas (Ed.), *Greek music in America* (pp. 87–104). University Press of Mississippi.

Horkheimer, M., & Adorno, T. W. (2002). *Dialectic of enlightenment*. Stanford, CA: Stanford University Press. Originally published in 1944. Edited by Gunzelin Schmid Noerr. Translated by Edmund Jephcott.

Ichikawa, H. (1963). A preliminary conception of Zen social ethics. *Studies on Buddhism in India*, *11*(1), 359–348.

Ichikawa, H. (1965). Preliminary conception of Zen social ethics (continued). *Studies on Buddhism in India*, *13*(2), 844–836.

Kaplan, D. M., Mehl, M. R., Pace, T. W. W., Negi, L. T., Ozawa-de Silva, B., Lavelle, B. D., Sivilli, T., Williams, A., Comstock, T., Price, B., Medrano, V., Robbins, M. L., Cole, S. P., Edward Craighead, W., & Raison, C. L. (2022). Implications of a "null" randomized controlled trial of mindfulness and compassion interventions in healthy adults. *Mindfulness*, *13*, 1197–1213.

King, S. B. (2016). Through the eyes of Auschwitz and the killing fields: mutual learning between Engaged Buddhism and liberation theology. *Buddhist-Christian Studies*, *36*, 55–67.

Kowalewski, D., & Greil, A. L. (1990). Religion as opiate: Church and revolution in comparative structural perspective. *Journal of Church & State*, *32*, 511–527.

Laing, R. D. (1967). *The politics of experience*. London: Penguin.

Le Guin, U. K. (1974). *The dispossessed*. New York, NY: Harper & Row.

Lebedev, A. V., Lovden, M., Rosenthal, G., Feilding, A., Nutt, D. J., & Carhart-Harris, R. L. (2015). Finding the self by losing the self: Neural correlates of ego-dissolution under psilocybin. *Human Brain Mapping*, *36*, 3137–3153.

Levine, D. H. (1988). Assessing the impacts of liberation theology in Latin America. *The Review of Politics*, *50*, 241–263.

Ligotti, T. (2010). *The conspiracy against the human race: A contrivance of horror*. Hippocampus Press.

Long, A. (2019). Ernest Hemingway in Turkey: from the Quai at Smyrna to A farewell to arms. *The Hemingway Review*, *38*(2), 75–86.

Martín-Baró, I. (1996). *Writings for a liberation psychology*. Cambridge, MA: Harvard University Press. Translated by A. Aron and S. Corne.

Marx, K. (1843/1970). *Critique of Hegel's philosophy of right*. New York, NY: Oxford University Press. Translated by Joseph O'Malley.

Marx, K. (1859/1977). *A contribution to the critique of political economy*. Moscow: Progress Publishers.

Maslow, A. H., & Honigmann, J. J. (1970). Synergy: some notes of Ruth Benedict. *American Anthropologist*, *72*, 320–333.

Mavelli, L. (2016). Governing uncertainty in a secular age: Rationalities of violence, theodicy and torture. *Security Dialogue*, *47*, 117–132.

McCloud, S. (2007). Putting some class into religious studies: Resurrecting an important concept. *Journal of the American Academy of Religion*, *75*, 840–862.

McLaren, P., & Jandrić, P. (2018). Karl Marx and liberation theology: dialectical materialism and Christian spirituality in, against, and beyond contemporary capitalism. *tripleC*, *16*, 598–607.

Metzinger, T. (2003). *Being no one: The self-model theory of subjectivity*. Cambridge, MA: MIT Press.

Metzinger, T. (2017). Suffering, the cognitive scotoma. In K. Almqvist & A. Haag (Eds.), *The return of consciousness* (pp. 237–262). Stockholm: Axel and Margaret Ax:son Johnson Foundation.

Minkov, M., Welzel, C., & Schachner, M. (2020). Cultural evolution shifts the source of happiness from religion to subjective freedom. *Journal of Happiness Studies*, *21*, 2873.

Netherland, J., & Hansen, H. (2017). White opioids: Pharmaceutical race and the war on drugs that wasn't. *BioSocieties*, *12*, 217–238.

Nichols, S., Strohminger, N., Rai, A., & Garfield, J. (2018). Death and the self. *Cognitive Science*, *42*, 314–332. Suppl. 1.

Palgi, P., & Abramovitch, H. (1984). Death: a cross-cultural perspective. *Annual Review of Anthropology*, *13*, 385–417.

Parfit, D. (2011). *On what matters*. Oxford, UK: Oxford University Press.

Pickering, J. (2019). 'I do not exist': Pathologies of self among Western Buddhists. *Journal of Religion and Health*, *58*, 748–769.

Potorti, M. (2017). "Feeding the revolution": The Black Panther Party, hunger, and community survival. *Journal of African American Studies*, *21*, 85–110.

Richards, G. (1978). Conceptions of the self in Wittgenstein, Hume, and Buddhism: An analysis and comparison. *The Monist*, *61*, 42–55.

Roy, D. J. (2015). We all die unfinished: a meditation on existential suffering. *Journal of Palliative Care*, *31*, 3–4.

Scott, J. C. (2009). *The art of not being governed*. New Haven, CT: Yale University Press.

Sedikides, C., & Wildschut, T. (2019). The sociality of personal and collective nostalgia. *European Review of Social Psychology*, *30*, 123–173.

Siderits, M. (2007). *Buddhism as philosophy*. Indianapolis, IN: Hackett.

Solt, F., Habel, P., & Grant, J. T. (2011). Economic inequality, relative power, and religiosity. *Social Science Quarterly*, *92*, 447–465.

Sturm, D. (1990). Martin Luther King, Jr., as democratic socialist. *The Journal of Religious Ethics*, *18*(2), 79–105.

Toutge, T. (2018). A startling injustice in the Protestant Reformation: The contrasting theologies of Martin Luther and Thomas Müntzer as seen in their responses to the Reformation of the Common Man, 1525. *Obsculta*, *11*(1), 9–33.

Vinding, M. (2020). *Suffering-focused ethics: Defense and implications*. Copenhagen: Ratio Ethica.

Zapffe, P. W. (1933). The Last Messiah. *Janus*, *9*. In Philosophy Now, Issue 45 (2004). Translated from the Norwegian by G. R. Tangenes.

7 A World to Win

The status quo. False consciousness and alienation. Critical consciousness: coming to senses. Optimism of the will. Selves in society. Armed love. The last revolution.

> That things are "status quo" *is* the catastrophe.
>
> *On the Theory of Knowledge, Theory of Progress*
> — WALTER BENJAMIN

> "But if abstract science is a luxury and practice of medicine mere chicane; if law spells injustice, and mechanical invention is but the means of robbery; if the school, at variance with the wisdom of the 'practical man,' is sure to be overcome; and art without the revolutionary idea can only degenerate, what remains for me to do?"
>
> Well, I will tell you.
> A vast and most enthralling task; a work in which your actions will be in complete harmony with your conscience, an undertaking capable of rousing the noblest and most vigorous natures.
>
> "What work?" — I will now tell you.
>
> *An Appeal to the Young*
> — PËTR KROPOTKIN (1880)[1]

The Status Quo

"IT IS A BEAUTIFUL, SUNNY JUNE DAY here in Ithaca, in the great state of New York. On a fallen log in my back yard, a pileated woodpecker is having his lunch. A fawn whose mom parked it for the afternoon in the bushes behind the deck is napping. An overweight rabbit pretends to ignore the neighbor's sheepdog, while munching down a dandelion leaf, the narrow end first. Squirrels are pursuing each other all over the stand of white pine and look momentarily smug whenever they catch up with themselves.

"On days like this, I am always moved to see the pursuit of happiness, which the United States Declaration of Independence lists as one of my inalienable rights, as just the thing for me. After all, I already have my life (good for as long as it lasts) and liberty (as much of it as the physics of the universe and my place in society allow). I have even experienced happiness occasionally (as recently as today), so I am sure I'll know it when I catch it again. But will I know what to do with it?"

S. Edelman, *The Consciousness Revolutions*,
https://doi.org/10.1007/978-3-031-24012-6_7

These two paragraphs, borrowed from a 2012 book on happiness, are quoted here to help me make a point about our species' present predicament — a point that is extremely difficult to get across to those who do not already see it, in part because some people have a stake in not seeing it and others are conditioned not to.

Any disquisition on happiness, especially one that invokes the Declaration of Independence, tells as much about its author as it does about its subject matter, and the two quoted paragraphs are no exception. Let's see what we can guess here. First, the author is upper middle class: a homeowner, in possession not just of an apartment, but an actual house with a deck and a back yard (the assertive "my back yard" sounds incompatible with merely renting). Second, the author is likely a professor (who else would mix physics with politics?), and a tenured one at that: Ithaca is a small town with a big university; adjuncts rarely own houses, nor are they prone to gushing about being happy; and the resting-on-laurels vibe here is too strong for this piece to have been written by a pre-tenure faculty member. Third, the author is probably a white cis male, a safe guess, given how dominant this demographic is among tenured professors.

Let us now consider the nature of the author's consciousness of his privilege. On the one hand, he realizes that a person's "place in society" affects their happiness. On the other hand, he cares neither about the origins of the society in question — the United States of America, built on lands expropriated through genocide using the labor of enslaved people — nor about its contemporary qualities. I should know: those two paragraphs were written, in the summer of 2011, by me.[2]

What did the U.S. society look like in 2011? The picture that can be gleaned from the readily available official statistics (compiled by agencies with a vested interest in whitewashing over the inconvenient truths),[3] is grim:

- Poverty rate: 15%, or 46.2 million people;
  - among all school-age children: 23%.
  - among African-American children: 39%.
  - among Hispanic children: 35%.
  - among Native American children: 37%
- People without health insurance: 48.6 million.
- People who used an emergency shelter or transitional housing program during 2011: an estimated 1.5 million.
- Military spending: $665 billion.
- State and local police spending: $99 billion.
- Federal minimum wage: $7.25 per hour.[4]

As with any other aspect of reality, what these numbers mean to a person is a matter of their consciousness. As we have seen in earlier chapters, consciousness encompasses both the interpretation of the agent's state of affairs and the caring about it. Efficiency in behavior regulation requires that daily routine — the status quo, especially of others, whose well-being is of no immediate concern to the agent — be treated as such: pushed to the cognitive background and effectively ignored, rather than allowed to disturb the agent's conscience.

This automatic moment-to-moment backgrounding of the suffering of others is complemented in the U.S. culture by a media milieu in which the only disasters are natural, the only people to blame for poverty are the poor, the military heroically defends our freedoms and exports democracy, the police serve and protect the people, the celebrities are duly celebrated, the philanthropists are hard at work eradicating hunger and disease in the Third World, the melting of the polar ice caps is a concern only for polar bears, and the Memorial Day sale is only weeks away.[5]

In this cultural setting, it is easy for a banker, a war profiteer, a cop, or a tenured professor at an Ivy League school to entertain class consciousness that shuts out anything that may trouble their sleep.[6] But what about the great majority of the people — those who themselves suffer the indignities of hunger, disease, or poverty, and who could put an end to all this preventable suffering, if they were to awake and rise up? That they succumb to *false consciousness* in considering the "status quo" as acceptable, or in seeing themselves as powerless to challenge it, is, I believe, what Walter Benjamin had in mind when he called it the catastrophe.

False Consciousness and Alienation

False consciousness is a state of mind that brings about, and is marked by, *alienation*. You don't have to have studied sociology or read Marx to recognize the attitudes and emotions that define it.[7] Do you feel *powerless* to change the direction of your country: the perpetual war abroad and at home, the dispossession of the people, the despoiling of the commons, the destruction of nature, the disruption of the planet's ecosystem, all in the name of profit and personal enrichment of the oligarchs? Do you feel that your life has devolved into a *meaningless* pursuit of career and commodities? Do you feel that the society is disintegrating into bands of *normless* individuals, who behave as badly as they can get away with, in proportion to the power they hold over others? Do you feel that people seek sanity by wrapping themselves in mental cocoons that *isolate* them from emotional engagement with one another and with the goings on in the world? Do you feel *estranged* from your very self, having lost sight of the person you had aspired to become back when you were growing up?

Our alienation goes to the roots. The realization of this is the essential springboard for any serious reflection on any aspect of present inter-human life. Viewed from different perspectives, construed in different ways and expressed in different idioms, this realization unites men as diverse as Marx, Kierkegaard, Nietzsche, Freud, Heidegger, Tillich and Sartre.[8] ... We are born into a world where alienation awaits us. We are potentially men, but are in an alienated state, and this state is not simply a natural system. Alienation as our present destiny is achieved only by outrageous violence perpetrated by human beings on human beings.
— R. D. Laing (1967, p.11)

In his introduction to *The Politics of Experience*, R. D. Laing invokes estrangement as a synonym for alienation: "Humanity is estranged from its authentic possibilities." If that is the case, then the experience of self-estrangement may actually help, by connecting one's personal experience to the species' collective predicament. Without such a connection, it would feel weird to speak of lofty concepts such as "humanity" or "authentic life" (except perhaps in a book). Indeed, one may ask how *this* book's average reader's alienation measures up when compared to the experience of people who really have it bad: the refugees from war or climate disasters, the uncared-for elderly, the homeless, the terminally lonely, the impoverished parents who work several jobs only to watch their children go hungry.

The answer is that it probably doesn't. At the same time, becoming conscious of one's own state of alienation is the all-important obligatory first step on the road to liberation. This is what Laing may have meant by remarking, in the same passage, that "No one can begin to think, feel or act now except from the starting-point of his or her own alienation." Comparing the condition of self to that of others can help individuals appreciate the structural causes of personal conscious experience, notably suffering. This, in turn, can make them see that liberation must surpass any personal "self-discovery" of the kind promoted by the spiritual traditions that came up in chapter 6.[9] It must do so by bringing about a society-wide abolition of the oppressive structures that cause eliminable suffering — what Ernst Bloch called "the transcendence of alienation by its conversion into praxis."[10]

The need to expand and hone one's consciousness so as realize and eventually overcome alienation runs counter to people's habit, also discussed in chapter 6, of suppressing self-consciousness in an attempt to blunt the suffering that accompanies existence. Here too the structural conditions that define one's station in life present a great challenge. Rather than merely setting free the suppressed self-consciousness, alienated individuals must overcome their *false consciousness* that grows out of "class bias and ideological programming"; in the words of Fredric Jameson,[11]

> The most influential lesson of Marx — the one which ranges him alongside Freud and Nietzsche as one of the great negative diagnosticians of contemporary culture and social life — has, of course, rightly been taken to be the lesson of false consciousness, of class bias and ideological programming, the lesson of the structural limits of the values and attitudes of particular social classes, or in other words of the constitutive relationship between the praxis of such groups and what they conceptualize as value or desire and project in the form of culture.

Taking as an example the U.S. politics, we can see that culture is indeed behind the most blatant manifestations of false consciousness. For instance, when working class people vote against their economic interests in support of politicians who oppose universal healthcare or living wage, it is because such politicians have successfully recruited these voters to their side in the "culture wars" on issues such as crime and terrorism, gun control, religion, or patriotism.[12] Indeed, the very notion that voting for one of the two parties in the American "democracy" can make life better for the majority of voters is itself a cultural myth, carefully cultivated by the political establishment that is deeply invested in keeping the masses in their state of false consciousness.[13]

The "negative diagnosis" of false consciousness offered by Marx must be supplemented by an analysis of the factors that shape it — a project that necessitates calling on history and anthropology, the disciplines charged with studying the actual patterns of human values, desires, and culture, before attempting to theorize about them. In particular, the temptation to equate culture with power, which can be traced back to Marx, must be resisted. Once we realize that it is culture that defines power, rather than vice versa, an opening is created for challenging and resisting power by shaping cultural dynamics. Anthropologist Marshall Sahlins has remarked on just how difficult this task is:[14]

> Built into perception, endemic in the grammar, working in the habitus, [cultural] structure is the organization of conscious experience that is not itself consciously experienced. Not easy to figure.

Critical Consciousness: Coming to Senses

The cultural matrix in which we are embedded, or rather, trapped — the "status quo" of Benjamin, the "habitus" of Sahlins — is not frozen in time, nor is its persistence due to mere inertia. As emergent, complex, entropy-defying systems under evolutionary pressure, cultures, like life itself, endure insofar as they are actively maintained by the societies within which they arise. One such maintenance mechanism, which seems to be universal across all human societies, takes the form of the often subtle social expectations to conform. In modern statist societies, it is complemented by — and in fascist states fully integrated with — the assertive, uncompromising, and ultimately violent combination of legislation and law enforcement.[15]

The investment in the upkeep of cultures pays off because of the benefits they confer on those who partake in them. By serving as cognitive shortcuts, cultural practices increase the efficiency of behavioral control and decision-making. This is especially important in complex environments, whose complexity grows as cultural artifacts, tangible or intangible, are integrated into people's lives. Indeed, cumulative growth of shared knowledge that constitutes culture is a defining characteristic of our species, which enables all of our technology and science. Lastly, culture is both a unifier and a divider: it promotes cohesion within groups, while deepening the divisions among them, a process that underlies multi-level evolutionary selection.[16]

This latter dynamics is one of the obstacles in the path of progress towards better life for everyone. The principles of justice should not distinguish between persons, let alone between members of cultural or ethnic groups. Thus, insofar as group-level cultural processes contribute to inequality along any dimensions that bear on people's well-being or, even more importantly, suffering, these processes are normatively bad and must be resisted. Empirical utilitarian considerations lead to the same conclusion: it turns out that redistribution in fact increases subjective well-being among the rich and the poor alike.[17]

History teaches us that pretending that well-being is a zero-sum game — the "us against them" sentiments and their culmination in the extreme individualistic ethic of "me first" or "me against everyone else" — is a recipe for perpetuating eliminable suffering. History lessons are, however, notorious for going unheeded, especially when the power elites act to suppress them by promoting ignorance and false consciousness.[18] At the same time, normative philosophical arguments and experimental findings that favor universal solidarity can only be effective if they succeed in countering the cultural dynamics that sustain inter-group polarization and conflict.

For that, the polarizing dynamics itself, as well as the role of the power players who facilitate it and benefit from its perpetuation, must first be exposed as such — that is, brought to the fore of people's consciousness. Our species' predicament is akin to that of a person who desperately needs therapy, yet does not realize it, or else resists the very idea of the need to change. If this analogy is valid, it offers a reason for hope. In therapy, the most important step is perceiving that one needs help and committing to work — and it usually *is* hard work — on bettering one's condition. Likewise, in changing cultural dynamics to the better, a conscious recognition of the need for such a change is key.

This may sound like a call to promulgate class consciousness, a notion that has become very familiar since its introduction by Marx. However, as Jameson has argued,[19] that is not quite what we need:

> [T]he notion of "class consciousness," as it is central in a certain Marxist tradition, rests on an unrigorous and figurative assimilation of the consciousness of the individual subject to the dynamics of groups. ... What is wanted here — and it is one of the most urgent tasks for Marxist theory today — is a whole new logic of collective dynamics, with categories that escape the taint of some mere application of terms drawn from individual experience (in that sense, even the concept of praxis remains a suspect one).

A good starting point for this project is the concept developed, and applied, by Paulo Freire, the great Brazilian educator, in the context of revolutionary teaching: *conscientização*, or coming-to-critical-consciousness.[20] In his work, Freire has famously distinguished between a "banking" model of education, in which the teacher attempts to deposit knowledge in the students' minds, and a critical-interactive approach, in which the teacher and the students are partners in generating and applying knowledge.

Freire's approach explicitly aims to engage with people's consciousness: "'Problem-posing' education, responding to the essence of consciousness — *intentionality* — rejects communiqué and embodies communication. It epitomizes the special characteristic of consciousness: being *conscious of*, not only as intent on objects but as turned in upon itself in a Jasperian 'split' — consciousness as consciousness *of* consciousness."[21] I may add that consciousness "turned upon itself" is perfectly poised to become critical. Its awakening can help us see in a new light both ourselves and the world that we create and inhabit. In the words of philosopher and political scientist Seyla Benhabib,[22]

> This philosophical anthropology can be viewed as a form of coming to one's senses morally, i.e. as a form of "Besinnung," a form of taking a hold of one's senses by grasping what it is to be human. What are some of the elements of such coming to one's senses? In the first place, an awareness of our natality as well as mortality, a cure for the sin, in St. Augustine's terms, of thinking that we are the ground of our being. We are not: we are fundamentally dependent creatures, born promiscuously to others like ourselves and radically dependent upon the good will and solidarity of others to become who we are. Furthermore, we are embodied creatures whose material needs must be satisfied by a constant engagement and metabolism with nature. This process of material engagement with the world is also one of world-constitution and world-creation.

Optimism of the Will

What kind of world *do* we want to live in? In their 1956 conversations (from which I already quoted back in chapter 4) about the possibility of a "new manifesto" for social change, Adorno and Horkheimer top up a critical assessment of the then state of global affairs with a prediction: "We can expect nothing more from mankind than a more or less worn-out version of the American system."[23] That, to put it mildly, is rather bleak.

Horkheimer: I do not believe that things will turn out well, but the idea that they might is of decisive importance.
— Adorno and Horkheimer (2010, p.43)[24]

Developing critical consciousness will not do us any good if it leads merely to nihilism and despair — or even just to an effective reconciliation with the steadily accelerating catastrophe that is the "status quo." Yes, our world is presently dominated by a culture of greed and egoism, which subjects billions of people to preventable suffering and is closing in on wrecking the ecosystem on which our species, and thousands of others, depend for survival. But people are still generally good: harried, depressed, and perhaps angry, but mostly open to cooperation and almost invariably kind to strangers in personal encounters.

Culture has been called by humanistic sociologist William du Bois "a description which we learn to see: a self-fulfilling prophecy which becomes reality." It is created by us, exists only by our leave, and can change if enough of us recognize that it must.[25] The role of critical consciousness in overseeing change is to guide the imagination towards the making of a truly better world. For that, we must regain a key freedom that has been lost when, as Graeber and Wengrow put it in *The Dawn of Everything*, "things went wrong." This is the freedom to imagine a radically different social arrangement from the one that dominates our existence at present, namely, the deadly partnership between capitalist economics and statist politics, which Chris Harman called "zombie capitalism":[26]

> The relationship between states and capitals are relationships between people, between those engaged in exploiting the mass of the population and those who control bodies of armed men. ... The state also has to provide mechanisms for integrating the mass of people into the system: on the one hand the coercive institutions that beat people into submission (police, secret police, prisons); on the other hand the integrative mechanisms that divert grievance into channels compatible with the system (parliamentary structures, frameworks for collective bargaining, reformist, conservative or fascist parties).

The alliance between capitalism and the state is precisely what went wrong with the ascendant culture on this planet.[27] This is why nothing less than an abolition of both will do to save it and us.

In contrast, half-measures invariably end up preserving the status quo. One familiar type of half-measure is economic reforms, which are particularly effective at strengthening the false consciousness that stifles any chance of a real change. The only people who are profiting from the think tank talk about "reimagining" capitalism and from the neoliberal state media extolling its citizens' "freedoms" are the usual suspects: on the one hand, the financiers, the corporate bosses, the rentiers, the tech moguls and monopolists, the "old money" families; and on the other, the politicians in their pay.[28]

The impossibility and, indeed, the mortal danger, of compromising with the capitalist state has been clear to many already a century and more ago. Rosa Luxemburg, in a pamphlet on *The Crisis in German Social Democracy*, written in prison in 1915, has framed the choice as that between socialism and barbarism:

> Today, we face the choice exactly as Friedrich Engels foresaw it a generation ago: either the triumph of imperialism and the collapse of all civilization as in ancient Rome, depopulation, desolation, degeneration — a great cemetery. Or the victory of socialism, that means the conscious active struggle of the international proletariat against imperialism and its method of war.[29]

Indeed, it was a policy of compromise — the collaboration between the German social democrats and the fascists pulling the strings of the Weimar Republic — that allowed the NSDAP to seize power in Germany in 1933.

Experiments in refusing to compromise with capital while keeping in place the state apparatus of oppression do not fare much better. The outcome in every past case — the military communism in Soviet Union in 1918-1921; the Castro regime in Cuba; the 1966-1976 cultural revolution period in China — has been the same: decades of general poverty and totalitarian control over all aspects of people's lives, eventually evolving into brutal state capitalism, of which present-day China is a prime example.[30]

It is unclear whether the complementary approach — doing away with the state while keeping capitalism — is at all viable, or, indeed, desirable. As I have argued in chapter 5, anarchism as a political system is not just historically and intuitively appealing: it is the only acceptable one, on a priori philosophical grounds.[31] However, large-scale historical examples of decentralized or stateless societies all predate the emergence of capitalism, which, moreover, was instrumental in the conquest and violent destruction of those societies.[32] With capitalism, which is antithetical to decentralization, in place and deeply entrenched, dismantling the state can only make the present corporate reign more fractured and even more violent. Anarcho-capitalism has about as much going for it as warlordism.

The remaining alternative is *anarcho-communism* — roughly, a system that combines the political principle "No gods, no masters" with the economic one "From each according to their abilities, to each according to their needs."[33] Given that such a system would spell the end of the rule of all political and economic power-holders, the project of preventing it from taking hold has the rare quality of being able to unify everyone on the traditional political spectrum, liberals and fascists alike.

Historically, grassroots attempts at combining political anarchism with socialist or communist economics have always come under relentless attack. In the nascent USSR, the anarchist movement of Nestor Makhno that operated in Ukraine has been suppressed by a temporary joint effort of Trotsky's Red Army and Denikin's and Wrangel's Whites, while in Moscow the Bolshevik secret police arrested and summarily executed hundreds of anarchist activists. In China between the revolutions of 1911 and 1949, both the communists and the nationalist Kuomintang acted against the anarchists. In Spain, the fascist rebellion led by Franco against the anarcho-syndicalist Republican government received critical support both from Hitler's Nazi Germany and from Stalin's "communist" USSR.[34]

Nor are present-day experiments in political anarchy safe from interference. The EZLN territory in Chiapas and similar indigenous enclaves in Oaxaca and Michoacán are under constant economic and military pressure from the Mexican state. The free Northern Kurdistan or Rojava, which effectively ended the ISIS rule in the region, is besieged by the government troops of fascist Turkey, cheered on by Russia and the U.S.[35]

In view of its presently precarious hold on geographical reality, anarcho-communism is thus literally utopian (of no place) in that it exists mainly in the minds of its adherents. Since being coined by Thomas More, the word "utopia" has, however, acquired an additional meaning: that of a noble undertaking, but an impossible one, as it would require a fundamental remaking of human nature.[36] It is therefore especially important to realize that in this sense, anarchism is, or should be, anything but utopian; as David Graeber noted in one of his last interviews, it is "deeply accepting of people as they are."[37]

Literary utopias do, however, vary in how thorough a reshaping of human nature they assume. On this scale, the prize for conceiving the most extreme makeover should probably go to Youwei Kang, an advisor to the Guangxu Emperor during the Hundred Days' Reform in late Qing China. Kang's 1902 blueprint for *Datong* (world-spanning "Great Harmony") proposed to transform even "the blackest African negroes [into] white persons."[38]

A stark contrast to this horror is the post-revolution world imagined by William Morris, graphic designer, writer, and socialist, in *News from Nowhere, or An Epoch of Rest*.[39] In that book, England has returned to its pastoral roots, except it is ruled by nobody: rural communities exist in peaceful anarchy; industry is small-scale, local, and driven by needs, not profit; and even London is transfigured into a kind of Elysium.

What one finds hard to believe about this story is not its conception of the new society, or its assumed appeal, but rather the feasibility of building it without interference from capitalist states, which have always been eager and capable of smothering any socialist initiative before it can take hold. Morris was well aware of this problem: his England, protected by the Channel from immediate interference from its neighbors, undergoes a quick revolution — which starts with a general strike[40] — and becomes such a shining example of socialism that the rest of the world follows suit.

A general strike is also how revolution starts in the world of Ursula K. Le Guin's *The Dispossessed*, subtitled "An ambiguous utopia" (which I will not attempt to summarize here – just read it).[41] Very importantly, a short story, *The Day Before the Revolution*,[42] not included in the book, puts "the" before "general strike":

> "The general strike," a voice, two voices, ten voices were saying in the room below, behind her.

The fifteen-page story makes it absolutely clear why: this is *not* merely another labor or politics-as-usual dispute — "Strikes for better wages, representation for women. Votes and wages — Power and Money, for the love of God!" — no, this is the strike with which the old order ends.

Selves in Society

The anarchist heroine of Le Guin's story, Odo, spends the day before the revolution ruminating about her life (she is seventy two), her friends (dead or imprisoned), and her cause:[43]

> But will you drag civilization down into the mud? cried the shocked decent people, later on, and she had tried for years to explain to them that if all you had was mud, then if you were God you made it into human beings, and if you were human you tried to make it into houses where human beings could live. But nobody who thought he was better than mud would understand.

What would it take for *us* to understand it? As a revolutionary-transformative principle, radical acceptance of human nature sounds like an oxymoron, yet it is in fact both logically compelling and supremely realistic. Whatever we, the people, want to be like in the long run, at this time we are what we are. Those destined to see the new world will not reach it unless *we* start working on it here and now. Perhaps if we work diligently enough, we ourselves will make it to the new world, hard as it may be to descry it from our present pit of despair.

Pessimistic takes on human nature are often driven by the best intentions. Thus, perceiving (correctly, I believe) that the greatest threat to our future arises from excesses of individualism, which makes people put the self first while the world burns, may motivate one to declare selfhood as such to be an obstacle to progress. That's how I interpret the position of Carlo Rovelli (mentioned in earlier chapters), which he supports by quotes[44] from fellow physicist Ernst Mach ("Das Ego ist unrettbar": the "I" cannot be saved) and from anarchist Alexander Bogdanov, who once wrote that "The individual is a bourgeois fetish."

Bogdanov's statement must, however, be considered in its original context, which suggests that the social dimensions of personal existence and consciousness can be valued without yet branding human nature as irredeemable:[45]

> The fourth fetish — the individual "I". A man considers himself autonomous, i.e., self-governing, standalone, separate being, an independent center of action, interests, yearnings, thoughts; he believes that in all that he is on his own. But that is not so. In actuality he is a product of the collective. His upbringing, his social milieu, his environment, his class, his group make him what he is, give form to his thinking and will. His actions depend on his upbringing and on social influences, that is, on the collective and its intentions; his thoughts are shaped by the same forces — upbringing, influences, that is, by the collective. There is nothing independent here. ... The personal, autonomous "I" is a very general fetish, which is rooted deeply in human nature; the core of the perversion is the placing of the individual in opposition to everyone else, and even to the entire world, whereas in practice he is a product of the society and the world.

In the years before, and for a short while after, the 1917 October revolution, Bogdanov had been a fellow party member and a key political rival of Lenin's. Eventually, however, his views got him sidelined (a rare fate, compared to that of other opponents of Bolshevism, who were put to death, unless they managed to emigrate). This is quite understandable. The Marxist-Leninist position has always been vanguardist in holding that during the revolution and its aftermath, the working class must accept the authority of the most ideologically advanced individuals. Of course, anarchists, who reject all authority, saw the "dictatorship of the proletariat," whether or not represented by a few individuals (as it was in the USSR and still is in China), as an anathema.[46]

It is possible to reject both capitalist-style individualism and "revolutionary vanguardism" (which is merely an incarnation of the classical hero worship) without advocating a dissolution of the self or its subjugation to the group. Can people be relied on to let go of the "fetish" of individualism for long enough to allow new and better social practices to take hold? One way to go about this matter is to follow Paul Goodman in insisting that "The issue is not whether people are 'good enough' for a particular type of society; rather it is a matter of developing the kind of social institutions that are most conducive to expanding the potentialities we have for intelligence, grace, sociability and freedom."[47]

Alternatively, we can accept that people are in fact already good enough, and have been all along, for the hundreds of thousands of years of our species' existence — a fact that somehow got lost in the conceptual upheaval of the European "Enlightenment," as noted by Graeber:[48]

> It is sometimes said that the central notion of modernism is that human beings are projects of self-creation. What I am arguing here is that we are indeed processes of creation, but that most of the creation is normally carried out by others. I am also arguing that almost all the most intense desires, passions, commitments and experiences in most people's lives — family dramas, sexual intrigue, educational accomplishment, honor and public recognition, one's hopes for one's children and grandchildren, one's dreams of posterity after one is dead — have revolved precisely around these processes of the mutual creation of human beings.

There is a normative implication to this point: if mutual creation is what we as humans do, it would all work better if we throw in some *love*.

Armed Love

> Of course our love is selfish. Almost all human love is bloody selfish. If one has anything to hang onto at all one clings to it relentlessly. We've tried to face it and to suffer it. To take refuge in love is an instinct and not a disreputable one.
> — Iris Murdoch (2001, p.46)[49]

As someone who has been brought up and educated in the individualistic Western tradition, I have always found it slightly embarrassing to hold forth on kinds of love, or, for that matter, friendship, other than the person-focused variety, a state aptly described by Elisabeth Vanderheiden as "a single soul dwelling in two bodies."[50] This made it very difficult for me to come to terms with the Greek concept of *agape* (especially in its Christian interpretation), or the kind of universal love inspired by its Dhārmic counterpart *ahiṃsā*, famously practiced by Mahatma Gandhi.

Ironically, it was a recent rereading of a very Western knightly romance that finally made me see the light (I think) with regard to the power of this latter kind of love.[51] Perhaps it was because the book, *The Well at the World's End*, had been written by a socialist and anarchist — William Morris. It describes a quest of two friends, and eventually lovers, Ralph and Ursula, for the water of a mythical well that brings "Quenching of Sorrow and Clearing of the Eyes" to those who drink of it.

At a critical stage of their adventure, Ralph and Ursula are instructed and aided by an old sage, who also delivers to them a warning:

> [S]ince the journey shall be yet with grievous toil and much peril, and shall try the very hearts within you, were ye as wise as Solomon and as mighty as Alexander, I will say this much unto you; that if ye love not the earth and the world with all your souls, and will not strive all ye may to be frank and happy therein, your toil and peril aforesaid shall win you no blessing but a curse.[52]

The precept to love "the earth and the world" resonates deeply with the anarchist vision of Kropotkin, whose *Mutual Aid* ends with these words: "And man is appealed to be guided in his acts, not merely by love, which is always personal, or at the best tribal, but by the perception of his oneness with each human being."[53]

> It is indeed necessary, however, that this love be an 'armed love,' the fighting love of those convinced of the right and the duty to fight, to denounce, and to announce.
> — Paulo Freire (1998, p.40)

Revolutionary and especially anarchist interpretations of *agape*, conceived as solidarity or comradeship, are arguably the best antidote to alienation, much of which is experienced as loneliness and disconnection from others.[54] Mere solidarity in (eliminable) suffering is, however, not enough: to even contemplate overthrowing their oppressors, the people must develop critical consciousness of their condition and its structural determinants.

As to whether or not love should be armed literally, and not just metaphorically... Because the monopoly on armed power on part of the state or of any group that seeks to boss over others is irreconcilable with real freedom, the anarchist answer is a resounding *yes*. Quoting Murray Bookchin, "The specific gravity of society, in short, must be shifted to its base — the armed people in permanent assembly."[55] Only then would the intent of the Creator in Alexander Pope's *Essay on Man*[56] be fulfilled:

> Thus God and Nature link'd the gen'ral frame,
> And bade Self-love and Social be the same.

The Last Revolution

Each successive consciousness revolution to date — the emergence, eons ago, of feelings that comprise selfless consciousness; the rise of local causal domains acting as minimal selves; of fully self-conscious subjectivity; of language and the explicit self-reflection it affords; and, finally, of societies of selves — has been about *more and more awareness*. An otherwise unremarkable planetary neighborhood, which started off as both lifeless and insensate, has broken out in a rash of sentience that made it a much more interesting place, populated with creatures that are capable of appreciating its new and wonderfully complex qualities. It would be a matter of cosmic irony if this exciting development were to be cut short, not by a stray comet or a gamma-ray burst, but by a failure of awareness on part of one species of such creatures. Locked into a cultural system that aggressively and violently defends itself against change, we are oblivious to our thralldom, to the self-inflicted nature of the suffering it brings, and to the irreparable damage it causes to our home planet.[57]

Not all of us are entirely unaware of this gruesome reality of our own making. While doomscrolling on social media, I have recently come across a cartoon that depicts an emergency box with a hammer and a sickle inside, labeled "in event of a capitalist crisis break glass."[58] There are two problems with this otherwise spot-on visual quip. First, we have reached the stage of capitalism when the crisis is not merely recurrent (as predicted by Marx),[59] but constant, making it difficult to decide when to commit to action and break the glass. Second, and more importantly, one needs a hammer to break the glass in the first place.

This second catch is precisely analogous to the state of affairs concerning education. Critical consciousness of the existence and the origins of the crisis and of the possibilities for its resolution can only develop in an educated people. This is why the educational system in a typical nation state ends up working to impart to its citizens — the "human capital" — just enough technical knowledge to be "competitive" (and often not even that), while keeping them ignorant about everything else.[60] The sure-fire way to prevent the people from taking the hammer and the sickle into their hands is to gaslight them about the emergency and, just to be on the safe side, hide the response box.

Those who manage, against all odds,[61] to become aware of the catastrophic dimensions of the status quo must still overcome strong psychological resistance to taking action, especially action that is as disruptive of everyone's daily routine as an abolition of the dominant social-economic order can be. The more resources we expend on getting where we are, the larger our emotional investment in the status quo, the more reluctant we are to give up the *sunk cost* and accept change.[62] On top of that, there is scary uncertainty in any change, and most of all in the singularity-ridden social dynamics of a revolution.

And yet, there is always hope.

Hope — especially educated hope, what Ernst Bloch called *docta spes* — has a peculiar dynamics of its own.[63] It is a mistake to think that hope, even if it is the last refuge of the powerless, is inherently passive. Remember the principle of predictive processing, which underlies self-evidencing and self-creation as discussed in chapter 3. In complex agentic systems such as ourselves, which represent both actual and desired states of affairs and are capable of allostasis or goal pursuit, predictive processing can take a desired state and proceed to actualize it. The imagined then becomes real — an act of *self-fulfilling prophecy*.

Popular examples of self-fulfilling prophecies typically involve initially false, and usually negative, beliefs that become a reality through social dynamics — as, for instance, when a groundless rumor of a bank's impending insolvency creates a run on withdrawals and causes a very real default, an example discussed at length by Robert Merton in his 1957 book *Social Theory and Social Structure*.[64] Merton's other examples are also all negative, having to do with the consequences of prejudice, out-group exclusion, educational inequality, and such.

The same dynamics of social belief fulfillment can, however, act in the opposite, positive direction. Instead of inventing specific examples here, I will just quote a line from a very well-known Chilean song written by Quilapayún — *el pueblo unido jamás será vencido* — and let your imagination do the rest.

Perhaps it is this kind of popular mobilization behind an initially empty slogan — a mere hope, as it were — that the young Marx intuited when he wrote, in a 1843 letter that Bloch quotes in *The Principle of Hope*:[65]

> Our motto must therefore be: reform of consciousness not through dogmas, but through the analysis of *mystical consciousness* which is still unclear to itself. It will then become apparent that the world has long possessed the dream of a matter, of which it must only possess the consciousness in order to possess it in reality.

A hopeful take on the self-fulfilling prophecy phenomenon is developed in William du Bois's 1983 thesis, *Love, synergy, and the magical – the foundations of humanistic sociology*:[66]

> If we seek to create a better self-fulfilling prophecy, then we gain a different understanding of "truth" and "reality." "Truth" becomes what we cannot readily change — the limits of our efforts to visualize a new world. Yet these "truths" themselves are subject to change almost daily. Thus the social disciplines are framed, in principle, by the idea that the potential for new visualization is inexhaustible. It is an effort in full opposition to any limits on our ability to change things: a heroic quest of the highest order.

The theory of predictive processing is the computational magic that can turn hope into action and aspirations into achievements. It complements the concept of self-fulfilling prophecy by offering an explicit, mechanistic account of how change can be brought about by a conscious exercise of the power of imagination.

The best future that we can imagine at this time may seem unattainable, but that must not prevent us from setting out, because our present condition is intolerable and inaction spells ruin. We may yet learn to imagine better as we go along,[67] as long as we heed the anarchist maxim of *prefiguration*: that the means we resort to in working towards change reflect, as best we can imagine them, the ends that we wish for.[68] All this likely means that there will be no "last revolution": future history should be open-ended. I shall therefore leave it to the reader's imagination and follow Walter Benjamin's advice by closing this chronicle of consciousness revolutions with his words[69] — "It is good to give materialist investigations a truncated ending."

> The place they go towards is a place even less imaginable to most of us than the city of happiness. I cannot describe it at all. It is possible that it does not exist. But they seem to know where they are going, the ones who walk away from Omelas.
> — Ursula K. Le Guin (1973)

❧ ❧ ❧

Notes

1. Walter Benjamin's status quo as catastrophe quote can be found in (Benjamin, 1999, p.473). The second epigraph is from (Kropotkin, 1909).

2. These two cringe-inducingly upbeat paragraphs open the seventh and last chapter of my 2012 book, *The Happiness of Pursuit*.

3. The poverty statistics are all from U.S. government sources for 2011. When consulting such sources, it must be remembered that official poverty figures

always undercount the actual cases: "The official U.S. poverty threshold is almost universally viewed as an antiquated and inaccurate measure" (Magnus, 2020, p.14). This problem is not unique to the U.S. (Lucci et al., 2018). For a comparative perspective, see Princeton economist Angus Deaton's New York Times op-ed, *The U.S. Can No Longer Hide From Its Deep Poverty Problem* (Deaton, 2018).

4. 73.9 million American workers age 16 and over were paid at hourly rates, representing 59.1 percent of all wage and salary workers. Among those paid by the hour, 1.7 million earned exactly the prevailing Federal minimum wage of $7.25 per hour. About 2.2 million had wages below the minimum. For context, see the previous note.

5. The following description of the corporate media in Latin America by Martín-Baró (1996, p.31) applies perfectly to the mainstream media all over the globe:

> Our countries live burdened by the lie of a prevailing discourse that denies, ignores, or disguises essential aspects of reality. The same 'cultural stranglehold' that day after day is imposed upon the people of Latin America by the mass media constitutes a reference point that has little bearing on the everyday experience of the majority of people, and even less on the poorer people. It goes along, deceptive and alienated, conforming to a fictional common sense that nurtures the structures of exploitation and conformist attitudes.

6. It is also easy for people, especially tenured professors, to rant about politics, because the U.S. political system (unlike, say, its Chinese or Russian counterparts) is perfectly confident in its immunity to mere criticism coming from niche sources. See (Edelman, 2020, ch.25), *Politics*.

7. The five aspects of alienation listed in the opening paragraph — powerlessness, meaninglessness, normlessness, isolation, and self-estrangement — are from (Seeman, 1959). Marx's early writings on the concept of alienation are analyzed in (Hoecker Drysdale, 1969). Appleton (2019) offers a useful historical overview (along with some unfortunate takes on identity politics). Morgan (2018) discusses alienation in the context of work. For this subject, see also the introduction from the forthcoming edition of *Mutual Aid: An Illuminated Factor of Evolution* by David Graeber and Andrej Grubačić, who write: "Kropotkin [in *Mutual Aid* (1902)] aimed to understand precisely what it was that an alienated worker had lost." Sawyer and Gampa (2020) document the empirical success of the alienation variable in predicting workers' subjective feelings.

8. Consider this footnote in (Laing, 1967, p.11):

> It is too late in the day now to go over the ground again covered by the thinkers of the last 150 years who have spelled out the nature of alienation, especially in relation to capitalism. For a succinct

summary, see Ernst Fischer, The Necessity of Art (London: Penguin Books, 1963), especially Chapter 3, 'Art and Capitalism'.

9. An interview with David Graeber (Graeber, 2012) contains a characteristically sharp and illuminating observation on alienation:

> The idea that alienation is a bad thing is a modernist problem. Most philosophical movements — and, by extension, social movements — actually embrace alienation. You're trying to achieve a state of alienation. That's the ideal if you're a Buddhist or an early Christian, for example; alienation is a sign that you understand something about the reality of the world.
> So perhaps what's new with modernity is that people feel they shouldn't be alienated.

10. The relevant passage appears in Bloch's essay *Karl Marx*, which is part of his book *The Principle of Hope* (Bloch, 1971, p.111):

> Hegel ended his *History of Philosophy* with a variation on a quotation from Virgil's Aeneid: *Tantae molis erat, se ipsam cognoscere mentem*; "What a burden for the mind to know itself." Marx always interpreted this burden as more than intellectual, even though he, together with Hegel, regarded the inscription at Delphi "Know thyself" as the theme of human history. For all that, he was far removed from any intention of defining self-knowledge (with the left-Hegelians) as the mere "philosophy of self-consciousness." Self-knowledge became non-contemplative and that of the working man, who sees himself in it both as the commodity which he becomes through alienation, and as a value-generating subject who transcends by revolutionary activity the commodity character forced upon him. This is the practical implementation of the Delphic admonition in Marx — the transcendence of alienation by its conversion into praxis.

In holding the attainment of right individual consciousness to be a prerequisite for collective action, I follow Bloch's analysis. According to Kellner and O'Hara (1976, pp.31-32),

> Bloch's stress on the importance of the subjective conditions of revolution can be seen as a reaction against deterministic-objectivistic-economistic versions of Marxism: "The exploding factor is the subjective factor of proletariat." ... "If the goals of a man fighting for higher wages do not include the disappearance of society that compels him to fight for higher wages at all, he will not get far in his fight for wages either."

11. The false consciousness quote is from *The Political Unconscious: Narrative as a Socially Symbolic Act* (Jameson, 1981, p.272).

12. Here are some entry points into the literature concerning: the war on terror (Robinson, 2017); fear as politics (Dutkievicz and Kazarinova, 2017;

Norton, 2018); the American culture of fear (Skoll, 2016); American populism (Brunello, 2019); culture wars and Covid-19 (Perry et al., 2020); and religion (Castle, 2019); and fascism (Fekete, 2022).

13. The U.S. voters' powerlessness has been quantified by Gilens and Page (2014).

14. The quote on the consciousness of culture is from (Sahlins, 1999a, p.413). For arguments to the effect that culture is not reducible to power, see (Sahlins, 1999b).

15. Prichard (2021, p.27) exposes the contradictions at the core of the concept of law enforcement:

> [I]f law is to have any independent normative force, and society is to be classless, then to use military or paramilitary power to enforce it would suggest two things. First that the normative force of law is undermined by the institutions which defend it, and secondly that it imposes and entrenches social conflict in the interests of those who define that order...

In this context, see also (Robinson and Barrera, 2012) on American fascism. For an empirical study of the "common is moral" heuristic and its role in the stability and change of norms, see (Lindström et al., 2018). Vallier (2017) describes a computational model of political stability. A complex systems perspective (with a focus on "democracies") can be found in (Wiesner et al., 2019). A concept related to that of political stability is societal resilience, discussed in (Folke et al., 2010) with a special focus on ecology and society; see also (Graves, 1991) on the dynamics of sociocultural evolution.

16. Nunn (2021) lists two functional benefits of culture — efficient decision-making in complex environments and the cumulative growth of knowledge that it affords — while noting the importance of group selection, for which also see (Odouard et al., 2021).

17. Watts and Hodgson (2019) provide a useful introductory overview of Rawlsian (Rawls, 1971) distributive theories of justice, noting that "Rawls is neither utilitarian nor Marxist; his theory of justice is Kantian and deontological and has been characterised as welfare state liberalism (Sandel, 1998), or egalitarian liberalism (Farrelly, 2004)." Unfortunately, Watts and Hodgson (2019) promote the classical liberal notion that equality of outcomes requires a curtailing of freedom:

> In contrast [to equality of opportunity], equality of outcome is more radical and egalitarian. It requires high degrees of political and economic intervention and coordination to bring about a situation where everyone has equal income, wealth and social well-being and ownership or resources. An implication of this position is that individual liberties, freedoms and differences may be subordinated in order to achieve the higher value of equality of outcome (Heywood, 1994).

Concerning the relationship between redistribution and happiness: correlational data from 136 countries are presented in (Kang and Rhee, 2021). (Cheung, 2018) summarizes his study, based on longitudinal data from Germany and other countries, as follows:

> [B]ased on 30 years of data from Germany and 24 years of data from across the world, increases in state- and national-level income redistribution over time were linked to greater life satisfaction, and this link was positive across individual, national, and cultural characteristics. The strong hypothesis of income redistribution — the idea that income redistribution benefits most people — gains empirical support from the current results. That is, income redistribution does not simply redistribute happiness. Rather, by redistributing income from the rich to the poor, citizens experience increases in life satisfaction across the board.

In a similar vein, Easterlin and O'Connor (2022) reported that "In Europe, differences among countries in the overall change in happiness since the early 1980s have been due chiefly to the generosity of welfare state programs — increasing happiness going with increasing generosity and declining happiness with declining generosity. This is the principal conclusion from a time-series study of 10 Northern, Western, and Southern European countries with the requisite data." A theoretical perspective that accounts for all these findings is that of relational happiness (Holmes and McKenzie, 2019). However that may be, the bottom line in this matter is that redistribution is a moral imperative (Hickel, 2019).

18. Concerning alienation and consciousness, Malherbe (2021b, p.10) writes that "... de-alienation entails a collective consciousness-raising process that seeks to understand how history finds form in an oppressive present so that we might change the present (see Freire, 1972; Montero, 2007)."

19. The Jameson quote on class consciousness is from (Jameson, 1981, p.284). It may be compared to this phrase from the work of the young Marx (1961, p.114): "Socialism is man's positive self-consciousness."

20. For the all-important concept of critical consciousness or *conscientização*, see (Freire, 1970). Ignacio Martín-Baró (1996, ch.2) discusses the parallel concept in Spanish, *conscientizacion*. A recent factor analysis of young people's political attitudes (Singh et al., 2020) has revealed the importance of critical consciousness.

21. The "consciousness of consciousness" quote is from (Freire, 1970, p.34).

22. Benhabib's (1996, p.54) views on embodiment and mutual creation are perfectly complemented by the discussion of embodied ethics in (Métais and Villalobos, 2021).

23. This statement appears in (Adorno and Horkheimer, 2010, p.36).

24. Horkheimer's acknowledgment here (Adorno and Horkheimer, 2010, p.43) of the possible importance of hope may be compared with its all-out defense by Ernst Bloch, whose views are mentioned throughout the present chapter.

25. Here is du Bois (1983, p.21ff) on the social construction of reality and the concept of self-fulfilling prophecy, which I shall highlight in the closing section of this chapter:

> Social reality is organized along the lines of the self-fulfilling prophecy and the definition of the situation (Thomas and Thomas, 1928; Merton, 1948). As Berger and Luckman (1966) have shown, social reality is constructed. Knowledge and belief are thus intimately connected with the actual construction of a reality. We create a world which mirrors our definitions of the situation. Society tends to be almost a "mass hypnosis" enacting those definitions as self-fulfilling prophecies.

26. The relevant passage in *The Dawn of Everything* is (Graeber and Wengrow, 2021, pp.132-133). The zombie capitalism quote is from (Harman, 2010, p.109-110).

27. For an argument concerning the alliance between capitalism and the state, see (Graeber, 2007a).

28. A representative piece of capitalist propaganda, designed to distract critics and defuse resistance, is the "stakeholder theory" (Freeman et al., 2016). For a devastating critique of the notion that capitalism can be reformed while retaining its defining tenets, see (Ci, 1999a,b). Graeber (2006) offers an invaluable historical-anthropological perspective. The political power of capitalism in the U.S. has been documented in (Gilens and Page, 2014). Possible ways of (radically) transforming capitalism have been discussed by Wright (2013) under the heading of "real utopias." For a commentary on the latter, (Brighouse, 2021).

29. The Rosa Luxemburg quote is from a 1915 pamphlet, written in prison, *The Crisis in German Social Democracy*. Interestingly, the phrase she attributes to Engels is actually from Kautsky ("The Erfurt Program: A Discussion of Fundamentals," 1892), as noted by Ian Angus (*The origin of Rosa Luxemburg's slogan 'socialism or barbarism'*, blog entry for October 21, 2014, on *Climate & Capitalism*, https://climateandcapitalism.com/2014/10/22/origin-rosa-luxemburgs-slogan-socialism-barbarism/).

30. That the institution of state is largely to blame becomes apparent from reading Scott (1998): *Seeing Like a State: How Certain Schemes to Improve the Human Condition Have Failed.* For a complementary critique of revolutionary anarchism (specifically in Europe in 1917-1923), see (Garner and Benclowicz, 2021).

31. To learn about anarchism, you can start with these sources: a short intro to the concept (Ward, 2004); its entry in the Stanford Encyclopedia of Philosophy (Fiala, 2021); its philosophical justification (Wolff, 1970); its defense by the some of its brightest proponents and practitioners (Goldman, 1911; Parsons,

2004); Black anarchism (Kom'boa Ervin, 1993); a geographical investigation (Scott, 2012); its anthropology (Graeber, 2004); the shape of anarchism in post-scarcity conditions (Bookchin, 1986); an ecological perspective (White and Kossoff, 2007).

32. See *The Art of Not Being Governed* by Scott (2009) and, of course, *The Dawn of Everything* by Graeber and Wengrow (2021).

33. In connection with the "Abilities to needs" principle, Spafford (2020, p.332) stresses the critical importance of the transfer of goods or services being *unconditional.*

34. For the definitive historical account of the Makhno movement, see (Arshinov, 2002); also (Garner and Benclowicz, 2021), and (Prichard, 2021, p.28). A discussion of the history of anarchism in China can be found in (Zarrow, 1990). Conlon (1986) summarizes the history of anarchist Spain in 1936-1938.

35. These and other contemporary anarchist alternatives to statist politics are discussed in (Prichard, 2021). For descriptions of ongoing anarchist praxis in Chiapas and Rojava, see, respectively, (Anderson and Springer, 2018; Maldonado-Villalpando et al., 2022) and (Dirik, 2022; Gerber and Brincat, 2021; Piccardi and Barca, 2022).

36. Kinna (2004) highlights the continued relevance of Thomas More's (1995) Utopia; Franzini Tibaldeo (2019) focuses on its anthropological significance and its power to fire the imagination.

37. Here is the exchange between Graeber and Nika Dubrovsky, one of the participants in the discussion (Graeber, 2020, p.79):

> ND: So in this way anarchism is if anything anti-utopian?
> DG: Yes. It is deeply accepting of people as they are.

38. In *Ta T'ung shu* or the Book of Great Harmony, Kang (1958, p.148) writes:

> Thus, in from seven hundred to a thousand years, the blackest African negroes will be transformed into white persons. By the time we have our One World, the people of all the earth will be of the same colour, the same appearance, the same size, and the same intelligence. This will be complete union — *ta t'ung* — of human races. (Those brown and black people whose natural dispositions are too evil, or who have diseases, will be given drugs to make them sterile...).

In this connection, see *Harmony: origin of totalitarianism or patron of pluralism?* (Li and Düring, 2020).

39. For those who wish to learn more about the author of *News from Nowhere* (Morris, 1890) and his politics, I highly recommend the works of E. P. Thompson (1959; 1977). A fascinating detail is found in (Kinna, 2004, p.745): "In June 1889, realizing that [Edward] Bellamy's book was being taken as a definitive description of socialist organization, [Morris] wrote a long and damning

review of it in *Commonweal*, dubbing its vision "State Communism" and the "very extreme of national centralization."

40. This and other aspects of the political views of William Morris are discussed in (Lloyd, 1977).

41. See (Jameson, 2004) on the politics of utopia in general and (Jameson, 1975, p.227) specifically on Le Guin's vision of the "ambiguous utopia" that she presents in (Le Guin, 1974b):

> Utopia is, in other words not a place in which humanity is freed from violence, but rather one in which it is released from the multiple determinisms (economic, political, social) of history itself: in which it settles its accounts with its ancient collective fatalisms, precisely in order to be free to do whatever it wants with its interpersonal relationships — whether for violence, love, hate, sex or whatever.

42. The general strike is mentioned on p.989 of (Le Guin, 1974a). For some background on collective action theory, see (DeMarrais and Earle, 2017).

43. This quote is from (Le Guin, 1974a, p.987).

44. See (Rovelli, 2021, p.184).

45. The translation of this passage (Bogdanov, 1920, pp.59-60) from the Russian is mine.

46. Sochor (1988, p.31) writes:

> [N]either Lenin nor Bogdanov had shared Marx's confidence in spontaneity as a way of engendering class consciousness. As Lenin stated, the spontaneous development of the labor movement leads to "its becoming subordinated to bourgeois ideology." Bogdanov and Lenin, however, each offered a different remedy. Bogdanov's was the conscious cultivation of proletarian values and beliefs; Lenin's was the strengthening of vanguard influence.

47. I have been unable to find the original source of the Paul Goodman quote, which I came across in (Honeywell, 2011, p.2), in Goodman's own publications. In any case, a similar view is expressed in (Li, 2014, p.907): "Personal autonomy is not free from socialization. The relevant question here is not how persons can be free of socialization, but rather what kind of socialization is conducive to personal autonomy."

48. Horkheimer and Adorno (2002) make it clear why "Enlightenment" belongs in scare quotes (see also note 28 on p. 112). Graeber's take on "mutual creation of human beings" (Graeber, 2006, p.75) may be compared to that of Marx, whose *Third thesis on Feuerbach* (Marx, 1969, para.6) states that "the essence of man is no abstraction inherent in each single individual" but rather the "ensemble of social relations."

49. It must be noted that it is one of Murdoch's characters, and not the author herself, who proclaims that "love is selfish."

50. See my essay in (Edelman, 2020, ch.18), *Love*, to which the following observation (du Bois, 1983, p.155) applies:

> [W]riting on love highlights one's own maladjustment. The literature on love is plagued by the fact that too often we learn more about the author's individual pathology than we do about love.

The "single soul, two bodies" phrase is part of the title of (Vanderheiden, 2021).

51. To be clear, I never did and still do not believe that love alone can save the world. To illustrate what I mean, here is a meme, whose original source I have been unable to ascertain, that came up in my Twitter feed a few months ago:

> *How I defeated fascism with the power of love.*
> Chapter 1: The Power of Love.
> The first step of my journey was realizing that it is impossible to defeat fascism with the power of love.
> Chapter 2: The Power of Incredible Violence.

52. Here is the rest of the Sage's words: "But rather I bid you to live in peace and patience without fear or hatred, and to succour the oppressed and love the lovely, and to be the friends of men" (Morris, 1896, Book 3, chapter 8), *They Come to the Sea of Molten Rocks.*

53. See Kropotkin's (1902) *Mutual Aid.* There is an interesting connection here to Morris's work: "Kropotkin, the great anarchist leader then in exile in London, wrote in his obituary of Morris [published in the 1896 issue of *Freedom*] that *News from Nowhere* 'was the most thoroughly and deeply anarchist conception of future society that had ever been written.'" (Lloyd, 1977, p.278). Concerning love and capitalism, Malherbe (2021a) writes that "Under capitalism, dispositional forms of being — such as love — are denigrated to having, with value assigned to that which is privately owned." Du Bois (1983, p.233) posits that "The capitalistic ideal of the nuclear self must give way to another conception: self in relationship."

54. See (Narayan, 2019): "[Black Panther leader Huey] Newton remarked on how 'revolutionary love' would mark the process of geographically and culturally disparate oppressed communities finding 'common cause' with one another and building forms of solidarity and camaraderie against the global reach of reactionary intercommunalism." This latter concept is introduced in (Newton, 1970). Perspectives on the concept of revolutionary love can be found in (Chabot, 2008; Crampton, 2019; du Bois, 1983; Freire, 1998; Newton, 1995; York, 2021).

55. The Bookchin quote is from *Post-Scarcity Anarchism* (Bookchin, 1986, pp.190-191).

56. See *Of the Nature and State of Man with Respect to Society*, (Pope, 1903, Epistle III). I first came across this passage from Pope in (Sahlins, 1969).

57. Wong and Bartlett (2022) offer an economic analysis that predicates the longevity of planetary civilizations on their collective "homeostatic awakening" — becoming aware of the perils of unlimited growth and consciously choosing a sustainable future.

58. The drawing of the emergency box (2012) is by cartoonist Philippe Cazaumayou.

59. Concerning the crises of capitalism, see for instance (Harman, 2010; Robinson and Barrera, 2012; Sinclair, 1976).

60. For a contrast between liberal and anarchist education, see (Suissa, 2001, p.629): "From the anarchist point of view, the question 'what should our society be like' is not merely 'overlapping with' questions about education, but logically *prior* to it" (emphasis in the original). Suissa (2010) offers a book-length treatment of these matters.

61. The odds here are against us because of the mismatch between the relatively fast time scale of social interactions and relatively slow processes of technological "progress" and climate change. This, of course, brings to mind the parable of the slowly boiled frogs. Only recently has the prospect of a climate catastrophe become the subject of an explicit scientific concern (Kemp et al., 2022).

62. The concept of sunk cost has been studied, for instance, by Feldman and Wong (2018); see also (Belisle et al., 2020; Polasky and Dampha, 2021; Sargisson and Schöner, 2020) on "hyperbolic discounting" of prospective future outcomes. An evolutionary psychology perspective can be found in (Palomo-Vélez and van Vugt, 2021); a cognitive psychological one in (Constantino et al., 2021).

63. See *The Principle of Hope* by Bloch (1986) and a commentary by Levitas (1990). Smith (2010) treats hope as a cosmological principle. Jobst (2021) places Bloch's ideas in the context of education. The bridge to action is hinted at in (Kellner and O'Hara, 1976, p.27):

> We are always more than we currently are, are always ahead of ourselves, always driven by potentialities which are still outstanding, which summon us to action and struggle. This dialectical fire within, burning through our needs, desires, dreams and potentialities, provides the basis of an anthropology of revolution.

64. Self-fulfilling prophecy is the subject of (Merton, 1957, XIII).

65. The quote from Marx in (Bloch, 1971, p.22) is from the same page of his 1843 letter to A. Ruge (Marx, 1992, p.209) as the one in the epigraph to chapter 4. Regarding the role of consciousness in self-fulfilling prophecy, see (Krishna, 1971, p.1104).

66. See (du Bois, 1983, p.104).

67. The idea of "imagining better as we go along" corresponds precisely to "the Zapatista principle *caminar preguntando* (asking, we walk)," highlighted by David Graeber (2007b):

> As one of the *maestros* put it at the *Escuelita*, autonomy "has no end." These words speak to a healthy awareness of the incompleteness of their ongoing experiment, and also to the way that building autonomy *can never* be considered perfect and complete. This dismisses any pretension of creating an ideal society that might, one day, claim to have reached its goal and to exist in its fullest expression.

For a brief account of Zapatismo as it reflects on this and other social desiderata, see (Anderson and Springer, 2018).

68. Concerning prefiguration, "a concept central to anarchist theory," Prichard (2021, p.26) writes:

> Prefiguration is a theory of ethics, and social and revolutionary agency, which posits that means and ends are mutually constituted and so must be consistent with one another.

Speaking of prefiguration, some immediately applicable ways of engaging in post-capitalist praxis are discussed in (Chatterton and Pusey, 2020). Chabot (2008, p.824) lists "five interrelated stages — cultural preparation, organization building, confrontation, political and economic non-cooperation, and parallel institutions" that can serve as "a good starting point for potential revolutionaries who seek concrete ways to promote active forms of agape."

69. The closing quote is from (Benjamin, 1999, N9a,2. p.473), *On the theory of knowledge, theory of progress.*

References

Adorno, T., & Horkheimer, M. (2010). Towards a new manifesto? *New Left Review*, *65*, 33–61. Translated by Rodney Livingstone.

Anderson, J. K., & Springer, N. J. (2018). Zapatismo as a resonant public pedagogy. *Latin American Perspectives*, *45*(3), 151–170.

Appleton, J. (2019). Self, society, alienation: From Marx to identity politics. In A. Kennedy & J. Panton (Eds.), *From self to selfie: A critique of contemporary forms of alienation*, Chapter 8 (pp. 129–146). London: Palgrave Macmillan.

Arshinov, P. (2002). *History of the Makhnovist movement 1918-1921*. London: Freedom Press. With an introduction by Voline and prefaces by Fredy Perlman and Nicolas Walter.

Belisle, J., Campbell, L., & Todd, M. (2020). Delaying the point of no return: Discounting evident in climate change policy preference. *Behavior and Social Issues*, *29*, 64–77.

Benhabib, S. (1996). Identity, perspective and narrative in Hannah Arendt's "Eichmann in Jerusalem". *History and Memory*, *8*(2), 35–59.

Benjamin, W. (1999). Convolute N [on the theory of knowledge, theory of progress]. In R. Tiedemann (Ed.), *The arcades project* (pp. 456–488). Cambridge, MA: The Belknap Press of Harvard University Press. Translated by Howard Eiland and Kevin McLaughlin.

Bloch, E. (1971). *On Karl Marx*. New York, NY: Herder and Herder. Original edition: *Über Karl Marx*, Suhrkamp Verlag (1968), from *Das Prinzip Hoffnung*, 3 vols. (*passim*), Suhrkamp Verlag (1959), Frankfurt am Main. Translated by John Maxwell. Available online at https://thecharnelhouse.org/wp-content/uploads/2018/05/Ernst-Bloch-On-Karl-Marx.pdf

Bloch, E. (1986). *The Principle of Hope*. Cambridge, MA: MIT Press. Originally published in German (1954) under the title *Das Prinzip Hoffnung*. Translated by Neville Plaice, Stephen Plaice, and Paul Knight.

Bogdanov, A. (1920). *Elements of proletarian culture in the development of the working class*. Moscow: Gosizdat. In Russian. Available online at http://escriptorium.univer.kharkov.ua/handle/1237075002/1511

Bookchin, M. (1986). *Post-scarcity anarchism*. Montreal, Quebec: Black Rose Books. Second edition, with a new introduction. Available online at https://libcom.org/files/Post-Scarcity%20Anarchism%20-%20Murray%20Bookchin.pdf

Brighouse, H. (2021). If you're a socialist you need the Real Utopias Project, whether you like it or not. In *Paper Presented at the Conference "Engaging Erik Wright: Between Class Analysis and Real Utopias," October 9, 2021*.

Brunello, A. R. (2019). Populist mobocracy, fear, and lies: the politics of American populism. *Inquiries Journal*, *11*(2), 1–8.

Castle, J. (2019). New fronts in the culture wars? Religion, partisanship, and polarization on religious liberty and transgender rights in the United States. *American Politics Research*, *47*(3), 650–679.

Chabot, S. (2008). Love and revolution. *Critical Sociology*, *34*(6), 803–828.

Chatterton, P., & Pusey, A. (2020). Beyond capitalist enclosure, commodification and alienation: Postcapitalist praxis as commons, social production and useful doing. *Progress in Human Geography*, *44*(1), 27–48.

Cheung, F. (2018). Income redistribution predicts greater life satisfaction across individual, national, and cultural characteristics. *Journal of Personality and Social Psychology*, *115*(5), 867–882.

Ci, J. (1999a). Disenchantment, desublimation, and demoralization: Some cultural conjunctions of capitalism. *New Literary History*, *30*(2), 295–324.

Ci, J. (1999b). Justice, freedom, and the moral bounds of capitalism. *Social Theory and Practice*, *25*(3), 409–438.

Conlon, E. (1986). *The Spanish civil war: Anarchism in action.* Workers Soldarity Movement. Available online at http://struggle.ws/spain/pam_intro.html

Constantino, S. M., Schlüter, M., Weber, E. U., & Wijermans, N. (2021). Cognition and behavior in context: a framework and theories to explain natural resource use decisions in social-ecological systems. *Sustainability Science*, *16*, 1651–1671.

Crampton, A. E. (2019). A literacy of armed love: confrontation and desire in aesthetic and critical projects. *Studies in Social Justice*, *13*(1), 94–117.

Deaton, A. (Jan. 24, 2018). *The U.S. can no longer hide from its deep poverty problem.* The New York Times Opinion, https://nyti.ms/2GdwOQr

DeMarrais, E., & Earle, T. (2017). Collective action theory and the dynamics of complex societies. *Annual Review of Anthropology*, *46*, 183–201.

Dirik, D. (2022). Stateless citizenship: 'Radical democracy as consciousness-raising' in the Rojava revolution. *Identities*, *29*(1), 27–44.

du Bois, W. D. (1983). *Love, Synergy, and the Magical — The Foundations of Humanistic Sociology.* Ph. D. thesis, William Penn College, Oskaloosa, Iowa.

Dutkievicz, P., & Kazarinova, D. B. (2017). Fear as politics. *Polis. Political Studies*, *4*, 8–21. In Russian.

Easterlin, R. A., & O'Connor, K. J. (2022). Explaining happiness trends in Europe. *Proceedings of the National Academy of Science*, *119*(37), e2210639119.

Edelman, S. (2012). *The happiness of pursuit.* New York, NY: Basic Books.

Edelman, S. (2020). *Life, death, and other inconvenient truths.* Cambridge, MA: MIT Press.

Fekete, L. (2022). Anti-fascism — A new horizon. *Race & Class*, *63*(4), 101–106.

Feldman, G., & Wong, K. F. E. (2018). When action-inaction framing leads to higher escalation of commitment: a new inaction-effect perspective on the sunk-cost fallacy. *Psychological Science*, *29*(4), 537–548.

Fiala, A. (2021). Anarchism. In E. N. Zalta (Ed.), *The Stanford encyclopedia of philosophy* (Winter 2021 ed.). Metaphysics Research Lab, Stanford University.

Folke, C., Carpenter, S. R., Walker, B., Scheffer, M., Chapin, T., & Rockström, J. (2010). Resilience thinking: Integrating resilience, adaptability and transformability. *Ecology and Society*, *15*(4), 20.

Franzini Tibaldeo, R. (2019). Thinking and behaving "otherwise": An anthropological enquiry into utopia, image and ethics. *Ethics & Bioethics (in Central Europe)*, *9*(1–2), 3–10.

Freeman, R. E., Parmar, B. L., & Martin, K. E. (2016). Responsible capitalism: Business for the 21st century. In D. Barton, D. Horváth, & M. Kipping (Eds.), *Re-imagining capitalism*, Chapter 10 (pp. 135–144). New York, NY: Oxford University Press.

Freire, P. (1970). *Pedagogy of the oppressed.* New York: Continuum. English translation 2011.

Freire, P. (1998). *Pedagogy of the heart.* New York, NY: Continuum.

Garner, J., & Benclowicz, J. (2021). A failure of praxis? European revolutionary anarchism in revolutionary situations, 1917-1923. *Left History*, *24*(1), 10–44.

Gerber, D., & Brincat, S. (2021). When Öcalan met Bookchin: The Kurdish freedom movement and the political theory of democratic confederalism. *Geopolitics*, *26*(4), 973–997.

Gilens, M., & Page, B. I. (2014). Testing theories of American politics: Elites, interest groups, and average citizens. *Perspectives on Politics*, *12*, 564–581.

Goldman, E. (1911). *Anarchism and other essays* (Second revised edition). New York: Mother Earth Publishing Association. Available online at https://theanarchistlibrary.org/library/emma-goldman-anarchism-and-other-essays

Graeber, D. (2004). *Fragments of an anarchist anthropology.* Chicago, IL: Prickly Paradigm Press. Available online at https://www.prickly-paradigm.com/titles/fragments-anarchist-anthropology.html

Graeber, D. (2006). Turning modes of production inside out or, Why capitalism is a transformation of slavery. *Critique of Anthropology*, *26*(1), 61–85.

Graeber, D. (2007a). Manners, deference, and private property: or, elements for a general theory of hierarchy. In *Possibilities: Essays on hierarchy, rebellion, and desire*, Chapter 1. AK Press. Available online at https://theanarchistlibrary.org/library/david-graeber-possibilities

Graeber, D. (2007b). *There never was a West, or, democracy emerges from the spaces in between* (pp. 329–374). Oakland, CA: AK Press. Available online at https://theanarchistlibrary.org/library/david-graeber-possibilities

Graeber, D. (2012). Another world: Michelle Kuo talks with David Graeber. *Artforum*, *50*(10).

Graeber, D. (2020). *Anarchy — in a manner of speaking.* Conversations with Mehdi Belhaj Kacem, Nika Dubrovsky, and Assia Turquier-Zauberman. Available online at https://theanarchistlibrary.org/library/david-graeber-anarchy-in-a-manner-of-speaking

Graeber, D., & Wengrow, D. (2021). *The dawn of everything: A new history of humanity.* London: Macmillan.

Graves, P. (1991). The persistence of memory: Dynamics of sociocultural evolution. *Cultural Dynamics*, *4*(3), 290–320.

Harman, C. (2010). *Zombie capitalism: Global crisis and the relevance of Marx*. Chicago, IL: Haymarket Books.

Hickel, J. (2019). The imperative of redistribution in an age of ecological overshoot: Human rights and global inequality. *Humanity: An International Journal of Human Rights, Humanitarianism, and Development*, *10*(3), 416–428.

Hoecker Drysdale, E. S. (1969). *The Theory of Alienation in the 1844 Manuscripts of Karl Marx*. Ph. D. thesis, Louisiana State University.

Holmes, M., & McKenzie, J. (2019). Relational happiness through recognition and redistribution: Emotion and inequality. *European Journal of Social Theory*, *22*(4), 439–457.

Honeywell, C. (2011). Paul Goodman: Finding an audience for anarchism in twentieth-century America. *Journal for the Study of Radicalism*, *5*(2), 1–33.

Horkheimer, M., & Adorno, T. W. (2002). *Dialectic of enlightenment*. Stanford, CA: Stanford University Press. Originally published in 1944. Edited by Gunzelin Schmid Noerr. Translated by Edmund Jephcott.

Jameson, F. (1975). World-reduction in Le Guin: The emergence of utopian narrative. *Science Fiction Studies*, *2*(3), 221–230.

Jameson, F. (1981). *The political unconscious: Narrative as a socially symbolic act*. Ithaca, NY: Cornell University Press.

Jameson, F. (2004). The politics of utopia. *New Left Review*, *25*(Jan Feb), 35–54.

Jobst, S. (2021). The principle of hope: Bloch's contribution to the praxiological understanding of education. *Conflict & Communication Online*, *20*(2).

Kang, H., & Rhee, D.-E. (2021). Does income (re)distribution matter for subjective well-being? Evidence from cross-country panel data. *Social Science Quarterly*, *102*(2), 706.

Kang, Y. (1958). *Ta T'ung shu: The one-world philosophy of K'ang Yu-wei*. London: G. Allen & Unwin. Written over 1884-1902. First published posthumously in 1935. Translated, with an introduction, by L. G. Thompson.

Kellner, D., & O'Hara, H. (1976). Utopia and Marxism in Ernst Bloch. *New German Critique*, *9*, 11–34.

Kemp, L., Xu, C., Depledge, J., Ebi, K. L., Gibbins, G., Kohler, T. A., Rockström, J., Scheffer, M., Schellnhuber, H. J., Steffen, W., & Lenton, T. M. (2022). Climate endgame: Exploring catastrophic climate change scenarios. *Proceedings of the National Academy of Science*, *119*(34), e2108146119.

Kinna, R. (2004). The relevance of Morris's utopia. *The European Legacy*, *9*(6), 739–750.

Kom'boa Ervin, L. (1979/1993). *Anarchism and the Black revolution. Pages from Prison No. 4* (2nd ed.). Philadelphia, PA: Monkeywrench Press. Available online at https://theanarchistlibrary.org/library/lorenzo-kom-boa-ervin-anarchism-and-the-black-revolution

Krishna, D. (1971). The self-fulfilling prophecy and the nature of society. *American Sociological Review, 36*(6), 1104–1107.

Kropotkin, P. (1909). *An appeal to the young*. Chicago, IL: Charles H. Kerr & Company. First appeared in French as "Aux Jeunes Gens" in *Le Révolté*, June 25; July 10; August 7, 21 (1880). Translated by H. M. Hyndman. This edition available online at http://dwardmac.pitzer.edu/Anarchist_Archives/kropotkin/appealtoyoung.html

Kropotkin, P. A. (1902). *Mutual aid: A factor of evolution*. London: W. Heinemann.

Laing, R. D. (1967). *The politics of experience*. London: Penguin.

Le Guin, U. K. (1973). The ones who walk away from Omelas. In R. Silverberg (Ed.), *New dimensions* (Vol. 3, pp. 1–8). Nelson Doubleday. Available online at http://sites.asiasociety.org/asia21summit/wp-content/uploads/2011/02/3.-Le-Guin-Ursula-The-Ones-Who-Walk-Away-From-Omelas.pdf

Le Guin, U. K. (1974a). The day before the revolution. *Galaxy Science Fiction Magazine, 35*(8), 17–30.

Le Guin, U. K. (1974b). *The dispossessed*. New York, NY: Harper & Row.

Levitas, R. (1990). Educated hope: Ernst Bloch on abstract and concrete utopia. *Utopian Studies, 1*(2), 13–26.

Li, C. (2014). The Confucian conception of freedom. *Philosophy East and West, 64*(4), 902–919. Special issue: Tenth East-West philosophers' conference, "Value and Values: Economics and Justice in an Age of Global Interdependence".

Li, C., & Düring, D. (2020). Harmony: origin of totalitarianism or patron of pluralism? *Journal of East-West Thought, 10*(2), 1–9.

Lindström, B., Jangard, S., Selbing, I., & Olsson, A. (2018). The role of a "common is moral" heuristic in the stability and change of moral norms. *Journal of Experimental Psychology: General, 147*(2), 228–242.

Lloyd, T. (1977). The politics of William Morris's "News from Nowhere". *Albion, 9*(3), 273–287.

Lucci, P., Bhatkal, T., & Khan, A. (2018). Are we underestimating urban poverty? *World Development, 103*, 297–310.

Magnus, A. (2020, June). *Poverty in the United States: An analysis of Its Measurement and the Long-term Social and Economic Costs*. M.A. thesis, University of Denver, Faculty of the College of Arts, Humanities, and Social Sciences.

Maldonado-Villalpando, E., Paneque-Gálvez, J., Demaria, F., & Napoletano, B. M. (2022). Grassroots innovation for the pluriverse: Evidence from Zapatismo and autonomous Zapatista education. *Sustainability Science*, *17*, 1301–1316.

Malherbe, N. (2021a). Considering love: Implications for critical political psychology. *New Ideas in Psychology*, *61*, 100851.

Malherbe, N. (2021b, 5). A psychopolitical interpretation of dealienation: Marxism, psychoanalysis, and liberation psychology. *Psychoanalysis, Culture & Society*, *26*(3), 263.

Martín-Baró, I. (1996). *Writings for a liberation psychology*. Cambridge, MA: Harvard University Press. Translated by A. Aron and S. Corne.

Marx, K. (1845/1969). Theses on Feuerbach. In *Marx & Engels selected works* (Vol. 1, pp. 13–15). Moscow, USSR: Progress Publishers.

Marx, K. (1961). Private property and communism. In *Economic and Philosophic Manuscripts of 1844*. Moscow: Foreign Languages Publishing House. Translated by Martin Milligan.

Marx, K. (1992). *Early writings*. New York, NY: Penguin. Introduced by Lucio Colletti. Translated by Rodney Livingstone and Gregor Benton.

Merton, R. K. (1957). *Social theory and social structure*. New York, NY: The Free Press.

Métais, F., & Villalobos, M. (2021). Embodied ethics: Levinas' gift for enactivism. *Phenomenology and the Cognitive Sciences*, *20*, 169–190.

More, T. (1995). *Utopia*. Cambridge: Cambridge University Press. Originally published in Latin in 1516. Edited by George M. Logan, Robert M. Adams and Clarence H. Miller.

Morgan, J. (2018). Species being in the twenty-first century. *Review of Political Economy*, *30*(3), 377–395.

Morris, W. (1890). *News from nowhere, or an epoch of rest*. London: Kelmscott Press. Available online at https://www.gutenberg.org/ebooks/3261

Morris, W. (1896). *The well at the world's end*. London: The Kelmscott Press.

Murdoch, I. (2001). *A fairly honorable defeat*. Penguin.

Narayan, J. (2019). Huey P. Newton's intercommunalism: an unacknowledged theory of empire. *Theory, Culture & Society*, *36*(3), 57–85.

Newton, H. P. (1970). *Intercommunalism: A higher level of consciousness*. Unpublished manuscript. Dr. Huey P. Newton Foundation Inc. collection, M0864. Dept. of Special Collections, Stanford University Libraries, Stanford, CA. Available online at https://viewpointmag.com/2018/06/11/intercommunalism-1974/

Newton, H. P. (1995). *To die for the people: The writings of Huey P. Newton*. New York, NY: Writers and Readers Pub.

Norton, H. (2017–2018). Government speech and the war on terror. *Fordham Law Review*, *86*, 543–563.

Nunn, N. (2021). History as evolution. In A. Bisin & G. Federico (Eds.), *Handbook of historical economics*, Chapter 3 (pp. 41–91). New York, NY: Academic Press.

Odouard, V. V., Smirnova, D., & Edelman, S. (2021). Polarize, catalyze, stabilize: conscience and the evolution of cooperation.

Palomo-Vélez, G., & van Vugt, M. (2021). The evolutionary psychology of climate change behaviors: Insights and applications. *Current Opinion in Psychology*, *42*, 54–59.

Parsons, L. E. (1905/2004). The principles of anarchism. In G. Ahrens (Ed.), *Lucy Parsons: Freedom, equality and solidarity*. Chicago, IL: Charles H. Kerr. Available online at https://theanarchistlibrary.org/library/lucy-e-parsons-the-principles-of-anarchism

Perry, S. L., Whitehead, A. L., & Grubbs, J. B. (2020). Culture wars and COVID-19 conduct: Christian nationalism, religiosity, and Americans' behavior during the coronavirus pandemic. *Journal for the Scientific Study of Religion*, *59*(3), 405–416.

Piccardi, E. G., & Barca, S. (2022). Jin-jiyan-azadi. Matristic culture and Democratic Confederalism in Rojava. *Sustainability Science*, *17*, 1273–1285.

Polasky, S., & Dampha, N. K. (2021). Discounting and global environmental change. *The Annual Review of Environment and Resources*, *46*, 691–717.

Pope, A. (1903). *The complete poetical works*. Boston: Houghton, Mifflin & Co. Edited by Henry W. Boynton.

Prichard, A. (2021). Taking the state back out: bringing anarchism back in. *International Politics Reviews*, *9*, 25–31.

Rawls, J. (1971). *A theory of justice*. Cambridge, MA: Harvard University Press.

Robinson, P. (2017). Learning from the Chilcot report: Propaganda, deception and the 'War on Terror'. *International Journal of Contemporary Iraqi Studies*, *11*(1 & 2), 47–73.

Robinson, W. I., & Barrera, M. (2012). Global capitalism and twenty-first century fascism: A US case study. *Race & Class*, *53*(3), 4–29.

Rovelli, C. (2021). *Helgoland: Making sense of the quantum revolution*. New York, NY: Penguin.

Sahlins, M. (1969). Economic anthropology and anthropological economics. *Social Science Information*, *8*(5), 13–63.

Sahlins, M. (1999a). Two or three things that I know about culture. *Journal of the Royal Anthropological Institute*, *5*(3), 399–421.

Sahlins, M. (1999b). What is anthropological enlightenment? Some lessons of the twentieth century. *Annual Review of Anthropology*, *28*, i–xxiii. Introduction to special issue.

Sargisson, R. J., & Schöner, B. V. (2020). Hyperbolic discounting with environmental outcomes across time, space, and probability. *The Psychological Record*, *70*, 515–527.

Sawyer, J., & Gampa, A. (2020). Work alienation and its gravediggers: Social class, class consciousness, and activism. *Journal of Social and Political Psychology*, *8*(1), 198–219.

Scott, J. C. (1998). *Seeing like a state: How certain schemes to improve the human condition have failed.* New Haven, CT: Yale University Press.

Scott, J. C. (2009). *The art of not being governed.* New Haven, CT: Yale University Press.

Scott, J. C. (2012). *Two cheers for anarchism: Six easy pieces on autonomy, dignity, and meaningful work and play*. Princeton, NJ: Princeton University Press.

Seeman, M. (1959). On the meaning of alienation. *American Sociological Review*, *24*(6), 783–791.

Sinclair, P. R. (1976). Fascism and crisis in capitalist society. *New German Critique*, *9*(Autumn), 87–112.

Singh, S., Javdani, S., Berezin, M. N., & Sichel, C. E. (2020). Factor structure of the critical consciousness scale in juvenile legal system-involved boys and girls. *Community Psychology.*, 1–17. https://doi.org/10.1002/jcop.22362

Skoll, G. R. (2016). *Globalization of American fear culture*. London: Palgrave Macmillan.

Smith, N. H. (2010). From the concept of hope to the principle of hope. In J. Horrigan & E. Wiltse (Eds.), *Hope against hope, Probing the boundaries* (pp. 3–22). Amsterdam: Rodopi.

Sochor, Z. A. (1988). *Revolution and culture: The Bogdanov-Lenin controversy*. Ithaca, NY: Cornell University Press.

Spafford, J. (2020). An anarchist interpretation of Marx's "Ability to Needs" principle. *The Journal of Value Inquiry*, *54*, 325–343.

Suissa, J. (2001). Anarchism, utopias and philosophy of education. *Journal of Philosophy of Education*, *35*(4), 627–646.

Suissa, J. (2010). *Anarchism and education: A philosophical perspective*. Oakland, CA: PM Press.

Thompson, E. P. (1959). *An English revolutionary*. A lecture to the Williams Morris Society, 1959. Reprinted in the Tribune Magazine, April 23, 2020. Available online at https://tribunemag.co.uk/2020/04/an-english-revolutionary

Thompson, E. P. (1977). *Wiliam Morris: Romantic to revolutionary.* New York, NY: Pantheon Books.

Vallier, K. (2017). Three concepts of political stability: an agent-based model. *Social Philosophy & Policy Foundation, 34*(1), 232–259.

Vanderheiden, E. (2021). "A friend? A single soul dwelling in two bodies." Friendship — a special kind of love. In C.-H. Mayer & E. Vanderheiden (Eds.), *International handbook of love: Transcultural and transdisciplinary perspectives*, Chapter 8 (pp. 135–153). Cham: Springer.

Ward, C. (2004). *Anarchism: A very short introduction.* New York, NY: Oxford University Press. Available online at https://theanarchistlibrary.org/mirror/c/cw/colin-ward-anarchism-a-very-short-introduction.a4.pdf

Watts, L., & Hodgson, D. (2019). *Distributive theories of justice*, Chapter 7 (pp. 117–132). Singapore: Springer.

White, D. F., & Kossoff, G. (2007). Anarchism, libertarianism and environmentalism: Anti-authoritarian thought and the search for self-organizing societies. In J. Pretty, A. S. Ball, T. Benton, J. S. Guivant, D. R. Lee, D. Orr, M. J. Pfeffer, & H. Ward (Eds.), *The SAGE handbook of environment and society*, Chapter 3 (pp. 50–65). SAGE Publications.

Wiesner, K., Birdi, A., Eliassi-Rad, T., Farrell, H., Garcia, D., Lewandowsky, S., Palacios, P., Ross, D., Sornette, D., & Thébault, K. (2019). Stability of democracies: a complex systems perspective. *European Journal of Physics, 40*, 014002.

Wolff, R. P. (1970). *In defense of anarchism.* New York, NY: Harper and Row. Available online at https://theanarchistlibrary.org/library/robert-paul-wolff-in-defense-of-anarchism.pdf

Wong, M. L., & Bartlett, S. (2022). Asymptotic burnout and homeostatic awakening: a possible solution to the Fermi paradox? *Journal of the Royal Society Interface, 19*, 20220029.

Wright, E. O. (2013). Transforming capitalism through real utopias. *American Sociological Review, 78*, 1–25.

York, M. (2021). Building a culture of revolutionary love: The politics of love in radical social transformation. In C.-H. Mayer & E. Vanderheiden (Eds.), *International handbook of love: Transcultural and transdisciplinary perspectives*, Chapter 10 (pp. 177–191). Cham: Springer.

Zarrow, P. (1990). *Anarchism and Chinese political culture.* New York: Columbia University Press.

Epilogue

The way out.

> If it keeps on rainin', levee's goin' to break.
>
> *When the Levee Breaks*
> — Kansas Joe McCoy & Memphis Minnie (1929)

The NOAA forecast, ordered via the little satellite comm device that I pack when hiking solo, was terse: a 40% chance of rain for that day, clearing up later in the week. There was no room for details in the 160-character message, and no means of getting more information: although my camping spot had a line of sight to a cell tower, high up on the crest of the Panamints and invisible to a naked eye, it was too far away for any reception.

December mornings can be cold, even in Death Valley. The first bit of rain came down as I was warming my hands on a cup of steaming tea and waiting for the freeze-dried eggs and bacon breakfast to reconstitute itself in its foil pocket. There is nothing quite like the scent of sagebrush after a rain in the Mojave desert and I was enjoying it to the fullest. Then the sky to the west, where the weather was coming from, cleared, and a rainbow painted itself for a few minutes against the blue-gray wall of the Panamints. I felt like Noah at Mount Ararat. It was going to be a great hiking day.

I was camped at the Red Amphitheater, at the end of a rough and rocky dirt track called the Hole in the Wall road, six miles from the nearest asphalt, which takes three quarters of an hour to drive, in good weather. On the old topo map, the track continues for at least six more miles, winding deep into the Funeral Mountains towards Pyramid Peak. My plan was to walk that road eastward, see what was to be seen, and enjoy the freedom of the desert.

The tracks of a car on a desert plain can remain visible for decades, but in the foothills it takes just one good flood to obliterate any hint of what may have been a perfectly passable jeep road. The arroyo I was walking into bore no sign of a track. The floods that had scoured it shaped heaps of gravel into winding ridges and rolled down colorful rocks from the mountains upstream. Many rotund cacti were in bloom. The cliff next to which I paused to have a bite to eat had fossil shells the size of my palm embedded in it. Then the rain returned.

I pulled my day pack's rain cover over it, put up the hood of my waterproof jacket, and resolved to press on. The rain intensified. A dark cloud was visibly creeping down from Pyramid Peak straight ahead. As I turned around, the now freezing rain, coming from the west, started pelting my face. I was

S. Edelman, *The Consciousness Revolutions*,
https://doi.org/10.1007/978-3-031-24012-6

walking as fast as the broken terrain allowed, holding both my trekking poles in one hand, then in the other, so that I could keep one hand at a time in a pocket and out of the gale.

By the time I got back to my tent (around half past four), my pants were soaked, and each of my hiking boots had a little puddle inside, but the rain seemed to be over. The scenery looked quite unreal: black, roiling rain clouds receding to the east, the red sun about to go down behind the Panamints to the west, and my little tent right there in the middle. I cooked and ate dinner, put my boots to dry on the dashboard of the rented Jeep, crawled into the tent, and went to sleep.

I woke up in complete darkness from the noise of rain drumming on the tent. I had slept only a couple of hours. The rain was getting mixed with sleet, but the tent was holding up well, and inside my sleeping bag it was positively cozy. Should I just sit it all out? I tried to estimate the amount of water that was being poured from the sky into the great mountain basin, at the mouth of which my car and my tent were sitting, just off the jeep track, on a little rise, shielded on the upstream side by a jutting clif.

After a while, I began to perceive my mental set — the warm glow of privilege of being well-equipped against the elements, the reluctance to leave the comfort of the tent — for what it was: a delusional cover-up of a sunk cost. My emotional investment in the warm bubble of the tent was keeping me from acting on a situation that was growing more and more dire by the minute. Should I stay put, I would most probably survive the coming of the *seyl*, the flash flood. Still, being stuck in the literal middle of nowhere once the jeep track is washed away and having to call for a rescue would be unbearably embarrassing. With that, came a change of consciousness. Hoping that the way out from the valley was still passable, I decided to act.

I must have broken a speed record tearing down the tent and stuffing it, with all its contents, into the back of the sleet-covered Jeep, in freezing rain, while wearing shorts and flip-flops (a later audit revealed, to my great satisfaction, that not a single tent stake was left behind in the process of bailing out). The Jeep's headlights showed that the two tracks — depressions in gravel made by the wheels — have each become a little river. As I started driving down the trail, these soon merged with other rivulets coming down from the hills, until the road, such as it was, could no longer be seen under a moving sheet of water, dragging along gravel, uprooted sagebrush, and increasingly large rocks.

I knew that under the water there was solid ground — sand or mud would have been so much worse — and the Jeep never went deeper in than the hubcaps, so I pushed on, steering by such memories of my previous passage up that road as I could muster. In the dark, I could only see straight ahead, so much of that navigation must have been guesswork, but I could not afford to stop: even with a four-wheel drive, it is much more difficult to start moving from a cold stop than not to stop in the first place. The critical milestone was going to be the Hole in the Wall — an opening, as wide as a couple of cars, where the only way out of the valley passes through a vertical sandstone barrier, four hundred feet high.

Although the Wall remained invisible in the darkness, I felt the passage, as the water became deeper, then receded a little as the valley broadened again. I kept driving through the torrent, aiming as much as I could in the direction that seemed downhill. I was lucky: the gravel and small rocks carried by the *seyl* must have been smoothing over the normally broken-up terrain, making it possible to drive anywhere through that valley, even though before the flood only the jeep track running down its side had been passable.

After a while, the torrent subsided and slowed down, then turned aside into what must have been the creek's habitual flood channel. In half a mile, I came to a little gravel barricade made by another stream, running at a right angle to my heading. Beyond it was the asphalt. I have reached California highway 190, the way out of Death Valley. As I would soon find out, the night's adventures were not yet over for me. But I was on solid ground again, and headed in the right direction — finally happy to return from the wilderness to a peopled country, and happy to have acted before it was too late.

❧ ❧ ❧

Index

A
active inference, 55, 60
  and sense of existence, 54
affect, 20, 27, 75, 76, 170
  negative, 42
agency, 40, 41, 60, 62, **79**, 80
  a physical view of, 46
  and causality, 62, 80
  and free will, 61
  unruly hand, 60
alcohol, effects of, 23
alertness, 8, 23
alienation, **185**, 194, 197, 198
allostasis, 56, 69, 195
altruism, 136, 148
  altruistic punishment, 137
amoeba, 8
  drunk, 23
anarchism, 114, **133**, 190, 193, 201
  and communism, 190
  and freedom, 194
  and the state, 106
  contemporary instances, 191
  historical instances, 190
  prefiguration, 196, 206
  role of consensus in, 116
anarcho-syndicalism, 71
anesthesia, 8, 10, 24
anthropic principle, 45
anti-natalism, 48
attention, 7, 54, 63
autopoiesis, 56, **78**, 82
awareness
  phenomenal, **8**, **23**
  self-, 7, 8
awareness, phenomenal, 9, 11, 13, 15–17, 20, 101, 194
  and representation, 13
  minimal, **15**, 16, 20

B
bacteria, 12, 16, 26, 38, 41, 57, 64, 66
  selfhood in, 80
blanket, Markov, 13, 25, 66, 81, 138
body ownership, 59, 60, 80
brain, 9–14, 23–25
  and consciousness, 22
  as virtual reality engine, 58
  computation, 15
  Markov blanket of, 66
  state, 11, 16, 21
  what it does, **10**
Buddhism, 24, 72, 83
  and alienation, 198
  and social engagement, 171
  and suffering, 170
  dependent arising, 71, 170
  emptiness, 44, 72, 77, 83, 170, 176
  nirvana, 170
  non-duality, 82
  Yogācāra, 24, 83

C
capitalism, 128, 132, 143, 144, **144**, 151
  and the state, 189
  as driver of alienation, 197
  as slavery, 152
  permanent crisis, 195
  reforming, 201

S. Edelman, *The Consciousness Revolutions*,
https://doi.org/10.1007/978-3-031-24012-6

state, 190
textbf, 106
caring, 3, 4, 8, 18–22, **22**, 27, 184
and self-evidencing, 55
and suffering, 42
causality, 36, 62
and agency, 62, 80
and existence, 44
and free will, 61
causal domains, 37
inferring, 64
local causal domain, 45
vs. randomness, 36
web of cause and effect, 36, 40, 61, 76
cell, 3, 12, 13, 25, 56
as a local causal domain, 38
class, 100, 130
consciousness, 169, 185, 188, 203
traitors, dogs as, 135
coffee, 71, 73
collective consciousness, 99, 100, 111
communism, 105
and the state, 190
computation, 10, 11, **11**, 24, 55
complexity of, 36
multiple realizability, 27
states and transitions, 11
computer, 40
as a dynamical system, 21
digital, 11
conscience, **137**, 138, 149, 184
evolution of, 149
consensus, 107, 116
control
and homeostasis, 56
feedback, 56, 68, 81
of behavior, 19, 26, 47, 63, 64, 67, 73, 96, 101, 187
political, 103, 140, 144, 150
predictive, 69
social, 149
cooperation, 96, 109, 133, 134, 136, 138, 139, 149, 189
conditional, 41, 136, 137, 139
evolution of, 116, 136, 137, 148
failing, 139
in wolves, 96
corporations, 105, 143, 144
psychopathy of, 151
cosmology, 44, 45
critical consciousness, **130**, **188**, 189, 194, 200
culture, **99**, 110, 138, 187
and false consciousness, 186
and minoritization, 101
and society, 99
change, 189
functional benefits of, 199

D

decision making, 19, 36, 138
democracy, 116, 142, 150, 171
dependent arising, 24, 44, 71, 72, 170, 171
depression, 75, 171
desert, 76, 127, 129, 131, 135, 217
determinism, 44, 80
dialectic, 75, 112, 132, 205
discernment, *see also* telling apart, 14, 21
intrinsic, 16
dissociated states, 108
dreaming, 78
drive, 18
Dynamical Emergence Theory of phenomenal awareness, 24, 26, 44
dynamical system, 16, 37, 41, 46, 68, 70, 72, 80–82
dynamics, 35, 37
metastable, 26

E

education, 98, 130, 195
for consciousness, **188**
emergence, 24, **37**, 44, 82
contextual, 46
intrinsic, 38
macrostates, 40
of causal domains, 38
of culture, 187
of life, 20
of self, 42
of space, 83
of the state, 132

of value, 18
vs. panpsychism, 24
emotions, 20–22, 27, 70, 74, 137, 185, 195
and sociality, 41
as computation, 35, 75
dimensions of, 43
in bacteria, 21
obligatory nature of, 35, 75, 170
enclosures, 142
Enlightenment, 112, 135, 193, 203
entropy, 64, 65, 81, 187
ethics, 136, *see also* morality, 137, 200, 206
and suffering, 177
individualistic, 187
Zen, 145
ethology, 82
evolution, 4, 7, 18, 27, 39
and existence, 78
and suffering, 43
Extended Synthesis, 134, 136
in cosmology, 45
of representations, 14, 26
of the self, 35
of valence, 65
simulations of, 137
existence, 53
sense of, 53

F
false consciousness, 101, 113, 132, 185
and culture, 186
overcome, 186
fascism, 187, 190
American, 199
defeating, 204
feelings, 4, 10, 22, 27, 35, 75, 166, 170
and existence, 8
flight simulator, total, 67, 69
foresight, 55, 65
free energy
and homeostasis, 82
and minimal selfhood, 82
and prediction error, 83
as bound on surprise, 66
principle, **66**
free will, **61**, 80
as illusion, 63
evolutionary roots, 63
vs. randomness, 61, 80
freedom, 128–130, **130**, 194, 199
and community, 131
dialectic of, 132
of individual in society, **138**

G
Good Regulator theorem, **64**, 67–70, 81, 111

H
habitus, 187
Handicap Principle, 18, 27
happiness, 42, 75, 116, 129, 183
and redistribution, 200
and regret, 83
relational, 200
un-, 75, 76
hard problem of consciousness, 82
higher-order theories of consciousness, 23
hive mind, 152
homeostasis, 54, 56, **56**, 65
and free energy, 82
homunculus, 81
honest signaling, 18, **27**
hope, 76, 106, **107**, 117, 139, 188, 201
educated, 195

I
identity, 147
personal and social, 113
virtual, 149
illusion, 79
of free will, 63
out of body experience, 60
room tilt, 58
rubber hand, 59
individualism, 128, 131, **132**, 133, 167, 187, 193
and collectivism, 138
and selfishness, 136
individuals, 39, 192
intention, 60, 94, 97
motor, 60
intentionality, 73, 94

L
language, 5
and cooperation, 96

and culture, 111
and persuasion, 107, 116
and self-consciousness, 101
as inherently social, 96, 98, 110
as magic, 94, 108
consciousness-shaping, 102
evolution of, 94, 95, **95**, 109
functions of, 95, 110
grammar, 97
metaphoric, 95
nouns vs. verbs, 43
propaganda, 105, 106, 128
role in experience, 74
use of, 93
learning, 39
credit assignment, 36, 42
exploration, 74
in bacteria, 12
Locality Principle, 39
reinforcement, 20, 27, 36, 39, 63, 82
tractability of, 37
levels
of analysis in science, 40, 46, 71
of consciousness, 16
liberation, 101, 170, 185
from suffering, 170
psychotherapy, 177
theology, 169, 175
life, 20, 56, 65
and death, 47
and mind, 56, 78
linguistics, 97, 109
local causal domain, 38, 39, 45, 56, 61, 66
Locality Principle for learning, 39, 45
love, 193, 204, 206
as antidote to alienation, 194
revolutionary, 204
the power of, 204

M

magic, **94**, 97, 98, 107
computational, 196
witchcraft, 108
meaning, **13**, 14, 21
in language, 93, 97, **97**
intrinsic, 14
meditation, 23
memory, 3, 13, 19, 76, 100
and free will, 62
autobiographical, 112
role in experience, 71
mental health, 171
mind-body problem, 10
Identity Theory, 10, 15, 24
model
-based reinforcement learning, 81
causal, 64
evidence, 67
generative probabilistic, 64, 65, 73, 81
of the world, 4, 19, 59, 66, 68
self-, 4, 63, 67, 68
used for prediction, 17
money, 141
dark, 142
mood, 70, 74
morality, 41, 138
as cooperation, 136
corporate, 143
culture dependence of, 148
science of, 136
movement
initiation, 60

O

oppression, 101, 102, 106, 114, 130
other minds, problem of, 15, 25

P

pain, 18, **19**, 20
panpsychism, 24
perspectivalness, 57
phenomenology
computational, 70
political consciousness, 134
positive psychology, 129
poverty, 104–106, 144, 163, 166, 168, 175, **184**, 197
prediction, 17
error, 70, 73, 74, 76
and valence, 75
foresight, 18
precision, 84

predictive processing, 55, 56, 60, 65, 76, 78, 195
  and illusions, 59
  and sense of existence, 77
  and thermodynamics, 63
progress, 132, 147
  moral, 136
  obstacles to, 187
  technological, 205
psychotherapy, 171
  liberation, 177

Q
qualia, 71, 72
quantum mechanics, 72, 80
  relational, 83

R
reading, 98, 99
reality, 9
  simulated, 108
  virtual, 4, **9**, 10, 57, 67
    dreaming, 78
recursion, 68, 69, 97
regret, 128
  anticipated, 138
relational
  happiness, 200
  nature of experience, 71, 83
  theories in physics, 72
relativity theory, special, 37
religion, 167, 174
  anthropology of, 169
  culture wars, 186
  theodicy, 169
    as praxis, 175
  theology
    liberation, 169
representation, **12**, 14
  in bacteria, 12, 26
  intrinsic, 13, 16, 26
  of dynamics, 19
  of events, 17
  of similarities, 71, 83
  of the world by the brain, 12
  relational, 71
  space
    trajectory through, 16, 35

S
self
  -awareness, 11, 57, 76
  -consciousness, 166
    and suffering, 166
  -creation, 56, 193, 195
  -estrangement, 185
  -evidencing, **54**, 55, 56, 60, 74, 82, 138, 195
    and valence, 55
  -interpretation, 21
  -knowledge, 83, 172
  -location, 9, 57
  -mastery, 83
  -reference, 4, 67, 68
  -regulation, 56
  and culture, 149
  and others, 41, 139
  minimal effective, 23, **39**, 40, 42, 43, 56, 61
    and agency, 60
  narrative, 100, 112
  the role in learning, 42
  the tragedy of, 42
  virtual, 39, 41, 43, 63, 75
  vs. person, 113
sensory substitution, 57
sentience, 4, 8, 41, 100, 107, 149, 166, 170, 194
similarity, 22
social consciousness, 138, 166, 167
socialism, 102, 114, 190
sociality, 41, 95
  eu-, 96
  pro-, 135, 137
  ultra-, 167
solitude, 129, 164
soteriology, 76
soul, 10, 101, 107, **117**
space
  multidimensional, 12
  representation, 12
  spanned by brain activity, 12
species being, 148
state (political entity), 105, 106, **132**, 142, 144, 190
  as instrument of oppression, 129
  resilience of, 140
state space
  trajectory through, 12, 73
states, phenomenal

privileged access to, 13
strange loop, 68
stream of consciousness, 63
stress, 74
subconscious, 8, 23
suffering, 4, **42**, 43, 46, 55, 75, 76, 139, 149
abolition of, 170
and music, 165
as religious virtue, 168
assuaged by religion, 168
causes of, 76
eliminable, **165**, 169, 173, 185, 187
evolutionary role of, 43, 48
medicalization of, 171
of others, 185
psychotherapy, 171
the role in learning, 42
surprise, 62, 64, 65, 73, 84
and free energy, 66
avoidance of, 74
estimation of, 66
expected, 74
minimization of, 65, 66
system, 13
subsystems as causal domains, 37

T
telling apart, 3, 14, 21
thermodynamics, 56, 65, 81
and prediction, 63
thought insertion, 60, 81
tool-making, 95
transparency, phenomenal, 75, 170
Turing test, 111

U
uncertainty, 55, 74, 84
and surprise, 65
utopia, 151, 191, 202, 203
real, 201

V
valence, 27, **55**, 70, 73
and prediction error, 75
and surprise, 65, 73
value, 18, 19
vision, 38, 45, 66
and self-location, 57
binocular, 9
disorders of, 58
eye movements, 58
lens inversion, 58

W
work, 146
writing, 97, 98
and culture, 110

Z
zombie, 15, 26, 170
capitalism, 144, 151, 189
corporations, 144
dance, 15

MIX
Papier aus verantwortungsvollen Quellen
Paper from responsible sources
FSC® C105338

If you have any concerns about our products,
you can contact us on
ProductSafety@springernature.com

In case Publisher is established outside the EU,
the EU authorized representative is:
Springer Nature Customer Service Center GmbH
Europaplatz 3, 69115 Heidelberg, Germany

Printed by Libri Plureos GmbH
in Hamburg, Germany